UP THE CISTERN

Up the Cistern

A Lavish Celebration of the Smallest Room

James Riddle

SPHERE BOOKS LIMITED
London and Sydney

First published in Great Britain by
Sphere Books Ltd 1984
30–32 Gray's Inn Road, London WC1X 8JL

Reprinted 1985 (twice)

TRADE
MARK

Set in Souvenir

Printed and bound in Great Britain by
Cox & Wyman Ltd, Reading

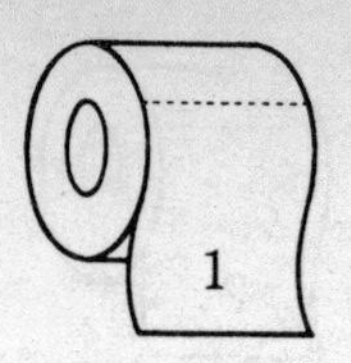

INTRODUCTION

'Small rooms and dwellings discipline the mind, large ones weaken it.'
Leonardo de Vinci

It is an astonishing and sobering fact that the room which we visit 5.6 times a day, which is the source of our earliest and most vivid childhood memories, and that gives relief and innocent pleasure to millions all over the world, is so often taken for granted.

Yet, as a glance through these pages will reveal, all human life is there. Kings have been born, have held court and died there. On a humbler level, all sorts of bizarre and usually rather unfortunate human incidents have taken place there.

And then, of course, there are the great mysteries. What is the world record for constipation? Why are senior diplomats allowed 28.8 toilet rolls a year and junior diplomats only 21.6? What has the Ayatollah Khomeini to say on the subject of toilet behaviour? Can head lice really be removed by application of urine mixed with fresh pigeon dung? Is it true that the French take no less than 87% of all medicine in the form of suppositories? And what *does* the Queen do about it when she's on tour?

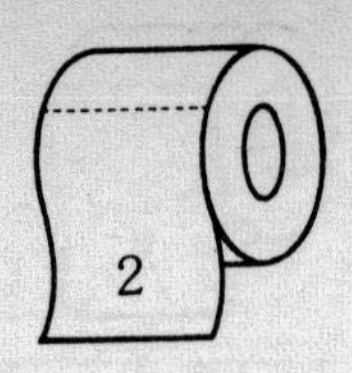

As well as answering these and other essential questions, I have included some help for those who find themselves ill at ease in these personal matters. Apart from providing no fewer than 100 ways of saying that you'd like to go and feed the goldfish, there are a number of basic etiquette tips and quizzes to keep you ahead in the toilet stakes.

But, above all, this is a celebration. With *Up the Cistern*, the privy goes public, the unmentionables become mentionable, the water closet comes out of the closet. Once again, we can sit in pride on our great heritage, rise refreshed, and flush with pride.

JAMES RIDDLE

February 1984

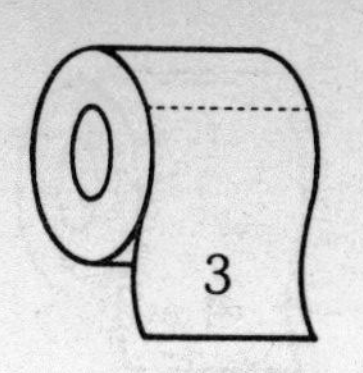

I

MOST INCONVENIENT

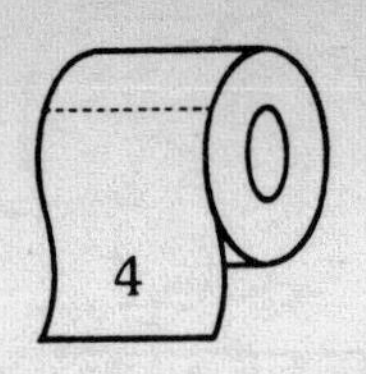

There was a young Royal Marine
Who tried to fart God Save the Queen.
When he reached the soprano
Out came the guano
And his breeches weren't fit to be seen.

★ ★ ★

There was an old lady of Chislehurst,
Who before she could pee had to whistle first.
One day in June
She forgot the tune
– and her bladder burst.

★ ★ ★

There was a young man of Loch Leven,
Who went for a walk about seven.
He fell into a pit,
That was brimful of shit,
And now the poor bugger's in heaven.

★ ★ ★

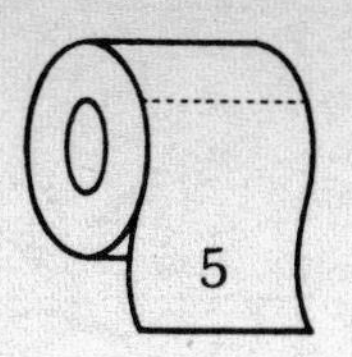

Are you still in there?

A statistical survey of usage

- **there are six billion lavatories with header tanks in the world**
- **in Britain lavatories are flushed a minimum of 3,600 times a second – and at peak periods up to 30,000 times a second**
- **the average number of times for going to the loo in a day is 5.6**
- **one in seven adults spends more than thirty minutes a day in the loo – and 3% of the population spend an hour or more**

- **The most popular pastimes in the loo are:**

  1. Thinking 21%
  2. Reading 20%
  3. Singing 6%
  4. Listening to the radio 6%
  5. Doing the crosswords 4%.

- **The most popular items in loos are:**

  1. Wall mirrors 54%
  2. Plants 22%
  3. Pictures/Cartoons 9%

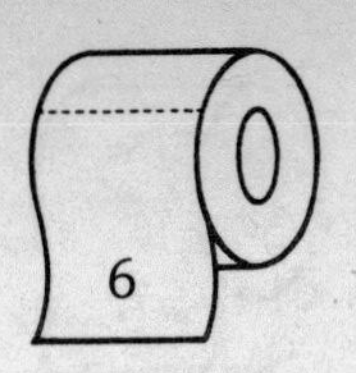

4. Chair/stool 8%
5. Magazines/newspapers 3%
6. Radio cassettes 2%

★ **The most desired fantasy extras for loos are:**

1. A heated seat
2. A built-in stereo system
3. Gold taps
4. Musical toilet roll holder

★ **43% of adults never lock the loo door**

★ **10% of all households still have an outside loo**

★ **53% of all loos are unheated**

★ **67% of all loos in Tyne Tees are unheated**

★ **98% of all homes use soft loo paper**

★ **The favourite colours for loo paper are:**

1. Blue
2. White
3. Green
4. Pink
5. Brown

- ★ In 67% of all cases, the colour and decoration of the loo is decided by a woman
- ★ 44% of all Londoners refer to it as 'the loo' as opposed to only 10% in the North East
- ★ More time is lost to industry through employees visiting the loo than through strikes

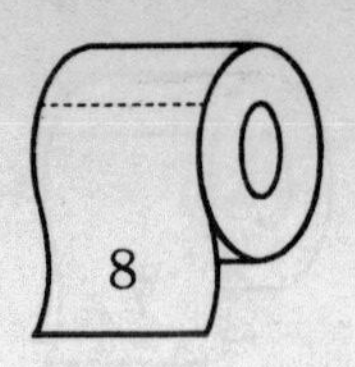

The Long Arm of the Law

'A man who stole a ballcock from an Abergavenny public toilet did it "because the Almighty told him it was to be done", the town's magistrate heard, but when asked what he was going to do with it, he replied that he did not know because God had not told him.'

Abergavenny Chronicle

'Toilet rolls were vanishing so fast at the police station that the senior officer ordered his men, "Use less or bring your own." The warning was part of a drive to cut the costs of the nick in Consett, County Durham, but after rumblings of discontent the order was withdrawn. A constable said, "We would have looked laughing stocks turning up for duty with truncheons in one hand and toilet rolls in the other." '

News of the World

'Police plan to ration toilet paper for prisoners following the discovery of two plaited ropes. The makeshift ropes were strong enough to hang a man or strangle someone.'

Daily Mirror

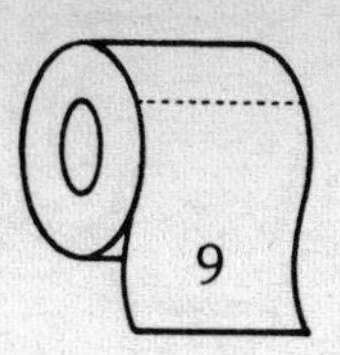

'A Nottinghamshire family, plagued for two years by a blocked-up lavatory, finally called in a plumber. He removed the trouble – a set of false teeth.'

People

When I was a wee wee tot
They took me from my wee wee cot
They put me on my wee wee pot
To see if I would wee or not
When they found that I would not
They took me from my wee wee pot
And put me in my wee wee cot
Where I would wee wee quite a lot

Traditional

The uncivil service . . .

It must be the longest running debate in government circles: the battle to get soft loo paper for civil servants. For the past decade, the staff have been calling for a gentler touch in the seat of power, but so far to no avail.

It's estimated that switching to soft toilet paper would double the Government's annual toilet paper bill of £750,000, and other costs caused by the introduction of soft two-ply, such as extra storage and handling, could add

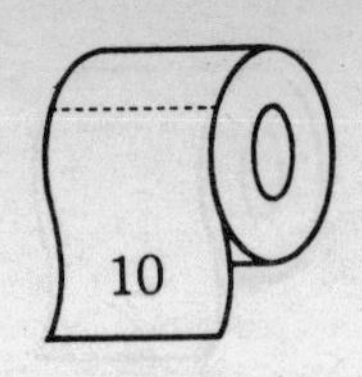

a further £200,000 to the bill . . . and there's no evidence Mrs Thatcher will be a soft touch.

The fact that some people have broken through has not alleviated the problem. There was an outcry when Sir Alasdair Milne, Director General of the BBC, had 'soft twin-ply' installed in his private lavatory, and the figures for overseas diplomats' spending on toilet items that are released by the Treasury show a tendency towards elitism.

Senior diplomat's allowance:	*Junior diplomat's allowance:*
57.6 bars of soap a year	43.2 bars of soap a year
28.8 toilet rolls a year	21.6 toilet rolls a year

When asked to explain these figures a Treasury official said, 'It will enable them to live the kind of life abroad that they would expect to live when they are at home.'

But one consolation for the Government is that prospective moles won't be tempted to Russia by their paper facilities. In Moscow the toilet tissue is in such short supply that many stores have worked out a system in which the customer pays a 2p a week subscription towards his toilet paper and gets a postcard when the paper arrives in stock.

Even then there's a snag. There's a shortage of postcards, too.

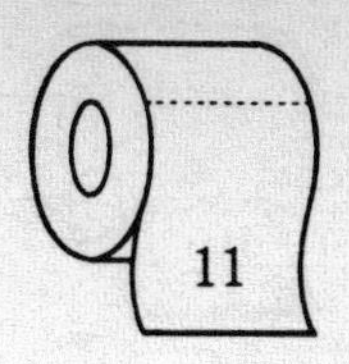

100 Loophemisms

'Tis needful that the most immodest word,
Be looked upon and learnt.'
William Shakespeare, *Henry IV, Part II*

The bard lived in more brutish times than our own, and nowadays we choose to wrap up the more intimate of our daily chores in a suitably obscure loophemism. Here is a selection of 100 choice phrases to help you through those difficult moments. Try them in the privacy of your own smallest room so that, the next time you're caught short while in company, you'll know how to ask to go to the – well, the, you know, that *place*.

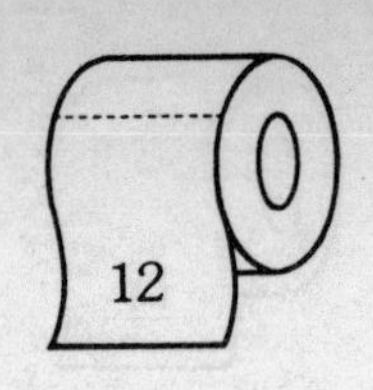

I would very much like to:

answer the call of nature

burn the grass
bury a Quaker

cash a cheque
check the plumbing
concentrate

do a job for myself
do a my word
do a puddle
do a rural
do an apple and pip
do my business
do my tables
do something no one else can
do for me
drown a mole

explain the chain

feel I could pick a daisy

go and feed the dog
go and feed the goldfish
go and see a man about a dog
go and shoot a dog
go and visit Sir Harry
go and visit Sir John
go and wash my hands
go to a place apart
go to check the telex
go to examine the back tyre
go to Jericho
go to mail a letter
go to my private office
go to powder my nose
go to see my auntie
go to shoot a lion
go to splash my boots
go to the bank
go to the bog
go to the comfort room
go to the diet
go to the euphemism
go to the House of Lords
go to the little boys' room
go to the little girls' room
go to the little house
go to the necessary house
go to the old soldiers' home
go to the rest room
go to the thinking place
go to tame the python
go to visit Mrs Jones
go to visit Mrs Murphy
go to visit my Aunt Jones

have a dicky diddle
have a dump
have a gypsy's kiss
have a hit and miss
have a Jimmy Riddle

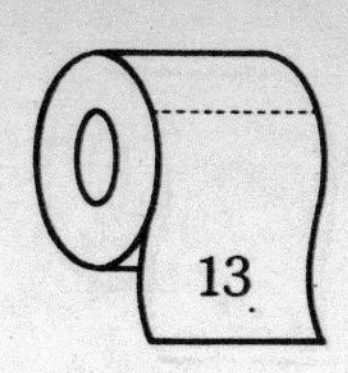

have a Johnny Bliss
have a leak
have a pony and trap
have a quiet time
have a rattle and hiss
have a slash
have a tom tit
have a you and me

make a call
make chamber music

obey the call of nature

pass a motion
plant a sweet pea
pluck a rose
point Alice at the Armitage
point Percy at the porcelain
pull out the one-eyed trouser snake
pump ship
put myself at ease

relieve myself

see if the horse has lost its blanket
see the geography of the house
shake hands with an old friend
shake hands with the bloke I once enlisted with
shake hands with my wife's best friend
shake the dew off a lily
shed a tear
sit on the throne
spend a penny
squeeze the lemon
strain my 'taters

tap a keg
tap a kidney
throw away some tea
turn the bike round

use the cloakroom
use the facilities

visit the snakes

wash my hands
water my dragon
water the lawn
water the nag
water the stock
wring my socks out

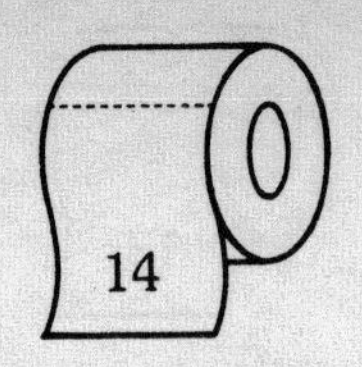

Loophemisms – the private sector

Some public schools boast of highly original names for their lavatories. A few examples:

Harrow — *Rears*
Rugby— *Topos*
Oundle — *Dykes*
Lancing — *Groves*
Marlborough — *Woods*
North Foreland Lodge — *Bogies*
Felstead — *Shankies*
Worth — *The Yard*

And the older universities are similarly obscure:

Balliol College, Oxford — *To visit Lady Periam*
Trinity College, Dublin — *Calfabias*
Christ's College, Cambridge — *To keep a fourth*
Trinity College, Cambridge — *Where's the rush matting?*

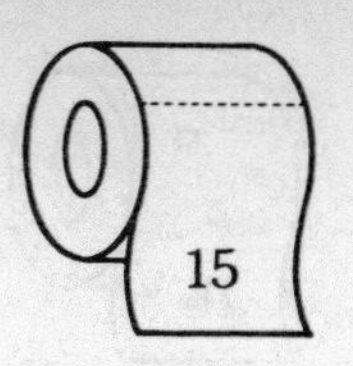

True or False?

The Queen has to have an unused toilet seat on any lavatory she is to use on an official visit.

(Although commonly held to be true, this is completely false. The Palace merely requests that a separate loo should be set aside for the use of Her Majesty alone.)

The first public lavatory in modern Britain was known as the 'place of the smells'.

(True. It was a two-seater built on the Isle of Skye for the relief of the men of the herring fleets and had its own unique flushing system that overhung the sea. The Gaelic name for the place was *rendha an fhaileadh* or the place of the smells.)

The inventor of the French stand-up lavatory was a penniless Russian emigré called Monsieur Uri Noir.

(False. Despite extensive research the name of the inventor is quite unknown.)

British Rail is obliged to provide toilet facilities in England on all stations where refreshments are sold.

(True.)

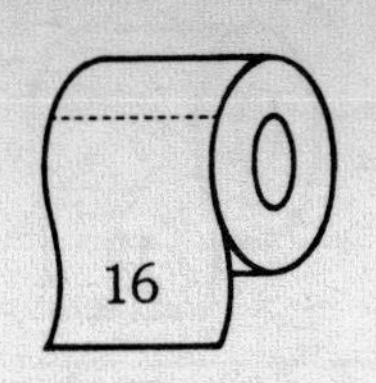

A cheque written on toilet tissue is legal tender.

(False. When a Mr Hancock of Swindon attempted to pay a driving fine on a toilet tissue cheque it was not allowed. Lord Hailsham, the Lord Chancellor, wrote, 'had he attempted to effect payment on a more suitable substance it might well have been accepted'.)

Recently, on prime-time television, Americans were told that you can expel the devil from within you by farting.

(True. In 1982, a sermon broadcast from Muncie, Indiana, told viewers that 'the escape of any gas from the body is a sign that the devil is making his exit'.)

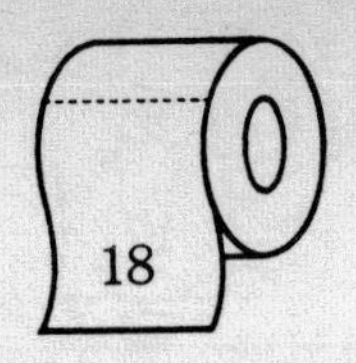

The cruellest cuts

In Torbay, Devon, cuts in government spending have led to the closing down of public conveniences. Postmen and milkmen are keeping their legs and their fingers crossed in the hope that the town will be provided with extra loos. At the moment though, many of them are being caught short by the shortage. One postman, flushed with embarrassment, ended up screaming in agony when, disturbed by a passer-by, he pulled up his zip too quickly. The wound turned septic and he was off work for a fortnight. Another postman was fined by the magistrates after police spotted him going in the bushes. He was so upset that he later quit his job.

Now Dr Marjorie Knobbs, a local GP, is appealing to householders to 'be neighbourly' and let delivery men use their toilets. She said, 'It can be a real problem for men who spend their time walking round estates where there are no public lavatories. I've seen several patients with bladder infections because they dare not drink much for fear of getting caught short.'

While the council denies there is any shortage of public lavatories, Dr Knobbs' campaign continues.

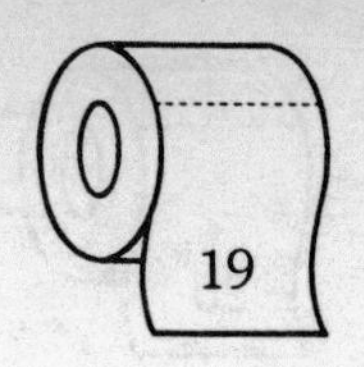

II

TO THE NECESSARY HOUSE

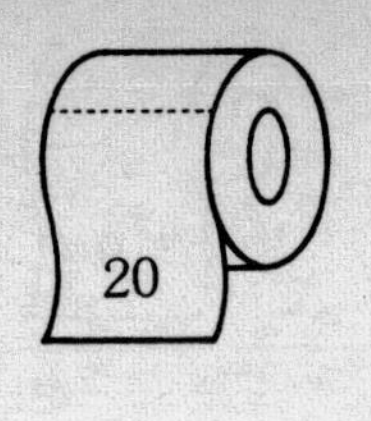
20

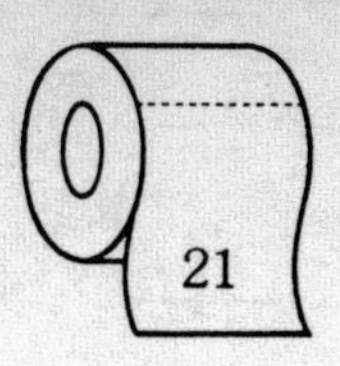

The parish commission at Roylette
Bought their vicar a pristine new toilet.
But he still voids his bowels
On a heap of old towels,
He's so damn reluctant to soil it.

A plumber who lived in East Dene
Designed an unusual latrine.
When seated you found
It emitted no sound,
When you rose it played God Save The Queen.

The Rajah of Afghanistan
Imported a Birmingham can
Which he set as a throne
On a great Buddha stone,
But he crapped out of doors like a man.

The first flush

The true originator of the smallest room was Sir John Harington, poet, inventor and courtier at the court of Queen Elizabeth I, whose godson he was. Born in 1561, Harington had a chequered career at court. His behaviour was, we are told, a little wild for the taste of the 'Virgin Queen' and he was expelled from London several times, on one occasion for translating lewd verses from the Italian and circulating them among the ladies of the court.

It was partly yet another attempt to win favour with the Queen and partly out of a genuinely fastidious sense of hygiene that, whilst banished to his family house at Kelston in Dorset, Harington came up with one of the greatest inventions of all time. Writing about it later, he was becomingly modest about the whole thing.

> *'The device was both first thought of, and discursed of, with as broad terms any belongs to it, in presence of six persons . . . of which I was so much the meanest . . . in a castle that I call the wonder of the west.'*

The Queen was apparently so impressed with the invention that she asked for one to be installed forthwith at Richmond Palace, and sent to her godson 'thanks for the invention'. Chained to the wall of the smallest room at Richmond Palace was a copy of Sir John Harington's book on the subject, pithily entitled *A New Discourse of a Stale Subject; Called the Metamorphosis of Ajax, Written by Misacmos to His Friend and Cousin Philostilpnos.*

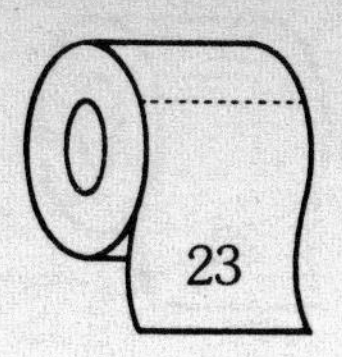

A reasonably witty celebration of the first proper loo, the book contains the lines:

A godly Father, sitting on a draught,
To do as need, and Nature hath us taught,
Mumbled, as was his manner, certain prayers:
And unto him, the Devil straight repairs,
And boldly to revile him he begins,
Alleging, that such prayers are deadly sins;
And that it proved he was devoid of grace,
To speake to God in so unfit a place.
The revered man, though at the first dismayed,
Yet strong in faith, thus to the Devil said:
Thou damned spirit, wicked, false, and lying,
Despairing thine own good and our annoying:
Each take his due, and me thou canst not hurt,
To God my prayer I meant, to thee the dirt.
Pure prayer to Him that high doth sit.
Down falls the filth, for fiends of hell more fit.

Sadly, Harington's efforts were largely in vain. It was over 175 years later that the first water closet was patented and went into production.

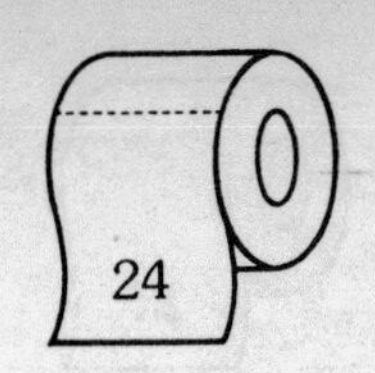

If you would hit the mark . . .

Moments spent in solitude in the smallest room can have a special kind of magic to them. Here are some poignant reflections on the subject from some of the world's great thinkers.

'If you would hit the mark, you must aim a little above it.'
H. W. Longfellow

'Only the actions of the just smell sweet and blossom in their dust.'
John Shirley

'Blow winds and crack you cheeks. Rage. Blow.'
William Shakespeare, *King Lear*

'The maxim of the British people is business as usual.'
Winston Churchill

'To business that we love we rise betime,
And go to't with delight.
William Shakespeare, *Anthony and Cleopatra*

'Wind bloweth where it listeth, and thou hearest the sound thereof but canst not tell whence it cometh.'
The Bible

'The evil that men do lives after them.'
William Shakespeare, *Julius Caesar*

'They shall not pass.'
Marshal Petain

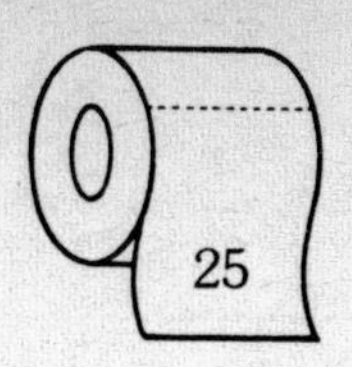

'For this relief much thanks.'
William Shakespeare, *Hamlet*

'Everyone places his good where he can.'
Voltaire

'All the beautiful sentiments in the world weigh less than a single lovely action.'
James Russell Lowell

'He that has done nothing, has known nothing.'
Carlyle

'The bitter and sweet come from the outside, the hard from within, from one's own efforts.'
Einstein

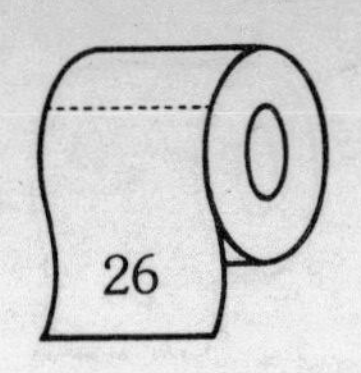

Bizarre Tales from The Small Room

There were emotional scenes when Mrs Alice Gilbert made her last appearance at the ladies public lavatory in St Paul's, Cray, where she had been the chief attendant for forty-five years.

Fifty well-wishers crowded into the lavatories for a farewell party, which ended with a song specially written for the occasion. It went:

'Good-bye, Mrs Loo,
We're sorry to be leaving you,
Boo-hoo!'

As the song ended, guests flushed all the lavatories in unison.

★ ★ ★

In the summer of 1976, Mr William Ferry of Plaistow was arrested for urinating in a public place. Protesting, Mr Ferry pointed to a tree nearby that had a 'GENTLEMEN' sign nailed to it.

'I always carry it with me in case of need,' he explained.

★ ★ ★

A campaign in the 1950s to encourage people to wash their hands after going to the lavatory got off to a bad start. *The Times* refused to accept their advertisement, saying that *Times* readers did not need reminding.

★ ★ ★

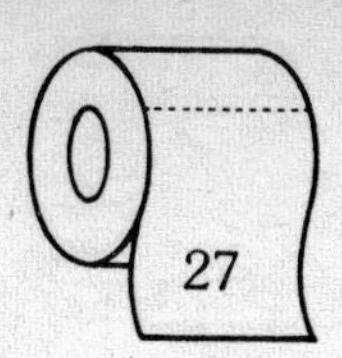

There was brisk bidding at an auction in Hollywood for some unusual antiques, including a lavatory seat once used by Adolf Hitler. The name of the lucky buyer was not revealed.

★ ★ ★

Neighbourly relations were strained in a row of terraced houses in Eastbourne when a Mrs Chick at No 40 called in the council sewage men to help clear her upstairs lavatory, which was badly blocked.

The council moved with great speed and soon a large sewage lorry with a team of operatives was attending to the crisis.

'They inserted their biggest pump into our convenience,' said Mrs Chick. 'One of the men switched it on. There was a tremendous rumble. Janet, my five year old, thought it was thunder. This rumble was followed by a tremendous whoosh. I nipped upstairs and flushed the convenience. It was working perfectly.'

But within moments of the sewage lorry driving off, Mrs Brown from No 42 next door was rushing down her neighbours' path in a state of hysteria. Instead of sucking, the council pump had blown and now the Brown's house was covered in effluent.

'Our house is ruined,' Mr Bob Brown told the *Daily Mail*. 'It's everywhere. The ceilings are in danger of collapsing. The lavatory pan was blown into the spare bedroom and the TV is all clogged up.'

★ ★ ★

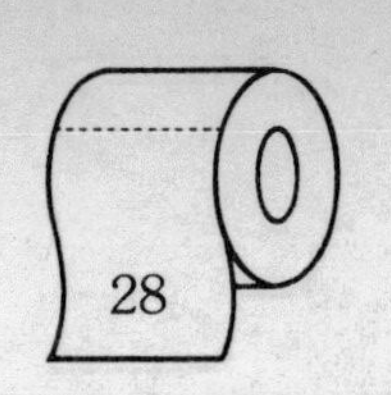
28

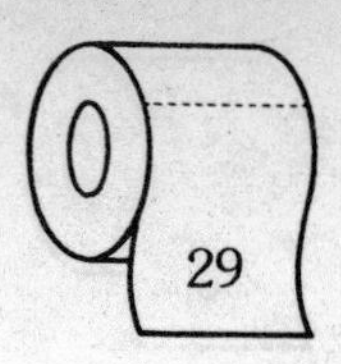

It was on the occasion of lady peers being admitted for the first time to the House of Lords in 1958, that a Naval Lord made the shortest speech recorded in the Upper Chamber. 'They would ruin the plumbing', he said.

★ ★ ★

A vet in Lictenvoorde, Holland, was called in to relieve Miss Utrecht, a prize cow, who was so inflated by her own gas that, according to the vet, 'she was almost three times her natural size'.

'As is usual in such cases, I inserted a large hose through Miss Utrecht's mouth and when it reached her stomach, applied a match to the free end. Much to my surprise, a flame roared forth which ignited the barn. In next to no time, the farm was a pile of ash.'

★ ★ ★

In 1971, a man in Surrey had a nasty shock while shaving. 'There was a loud swishing sound followed by a great thud,' he said.

When his wife went out to see what had happened, she found a massive lump of ice embedded in their lawn. The police were called in and took the ice block into custody.

It was then that someone realised that it must be a block of frozen urine jettisoned by a passing aircraft.

'We are very lucky it wasn't something worse,' said a police spokesman.

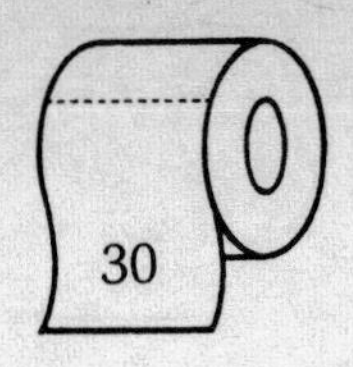

After each of her four marriages, Joan Crawford changed all the lavatory seats in her Hollywood mansion.

★ ★ ★

In 1967, an exceptionally fat woman was at the centre of a dramatic but embarrassing incident some 30,000 feet above the Atlantic.

She was sitting on a loo during a Pan Am flight when she found that she was inextricably wedged. Her bulk covered the entire seat and, when the plane reached 30,000 feet, extraordinary dynamic pressures built up, causing a suction effect between the woman and the loo.

She called for help and first one, then two, stewardesses attempted to lift her off the seat. When they failed, the pilot was summoned to lend a hand. He also failed.

Finally, the plane had to be taken down to 5,000 feet, where the lower pressure allowed the woman to be released like a cork out of a bottle.

Shortly afterwards, Pan Am announced the introduction of a new design of lavatory seat.

★ ★ ★

On all nine space shuttle flights, there has been one significant technical hitch – the space age toilet has broken down. The astronauts, including two women, have had to resort to good old plastic bags.

★ ★ ★

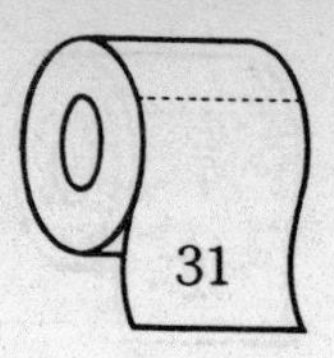

A senior member of the Government was caught short while walking home late at night and, thinking there was no one looking, relieved himself in Trafalgar Square. A policeman caught him in the act and when he asked the distinguished gentleman what he thought he was doing he was told briskly, *Necessitas non cogat legam* (Necessity knows no law). The policeman assumed he was a foreigner and arrested him on the spot.

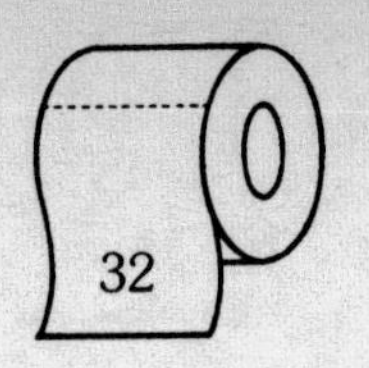

Beans, Beans,
Good for your heart,
The more you eat,
The more you fart,
The more you fart,
The more you eat,
The more you sit on the toilet seat.

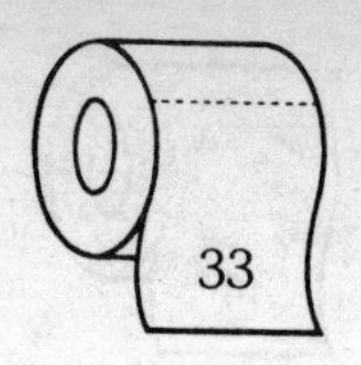

Are you ready for the Executive Toilet?

We all know that toilet-training affects every aspect of adult life from sexual preferences to the way you park your car. But did you know that your deepest personality traits are revealed in what psychiatrists refer to as TBS, Toilet Behaviour Syndrome?

Are you on your way to self-achievement? Or are you always on the outside, waiting for life's 'ENGAGED' sign to change to 'VACANT'? This simple personality test will reveal all.

1. Someone who is near and dear to you is always on the loo when you want to use it. Do you:

(a) put it down to your body rhythms being in harmony, which in its way is rather touching?

(b) quip light-heartedly that unless they hurry up, you'll have to go in their shoe?

(c) come on like Jack Nicholson in *The Shining*?

2. You have been on the loo ten minutes and nothing has happened. Do you:

(a) give up?

(b) turn to the next chapter, making a note to yourself that you should look into this F-Plan diet?

(c) decide that your body needs to be shown who's boss and stay there all day if necessary?

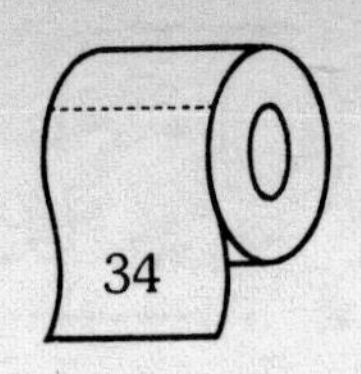

3. On taking up a new job you discover that you are barred from the executive toilet. Do you:

(a) feel grateful that you are allowed to go to the toilet at all at work, without deductions in pay?

(b) picket outside the loo with a 'Down with Toiletism!' banner?

(c) play the old cling film over the bowl trick on the managing director?

4. Your boss announces that the firm's traditional four and a half minute toilet allowance every morning has been suspended to boost productivity. Do you:

(a) make pathetic jokes about payment in lieu but put up with it?

(b) fart spectacularly during meetings?

(c) claim that four and a half minutes has always been a totally derisory management response to your members' quite legitimate aspirations and call for major industrial action unless the time is increased to thirty-seven minutes, plus a reasonable allowance for washing hands, adjusting clothing, finishing crosswords, etc?

5. While on holiday on a French campsite, you are obliged to use one of the unsavoury hole-in-the-ground toilets the French are so fond of. As you struggle with your clothing, your watch slips off and disappears down the hole. Do you:

(a) mutter 'C'est la vie' and resolve never to buy a Korean slip-on watch-strap again?

(b) fetch a fishing-net from the tent, telling yourself that improvisation is what camping holidays are all about?

(c) threaten the campsite manager with legal action under Clause 53(b) sub-section ix of the EEC Toilets Act unless he personally rolls up his sleeve and gets down to it?

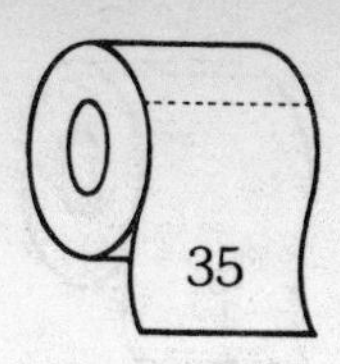

6. An attractive girl that you have been chatting up at a party joins the loo queue behind you. When you have finished, you find that your impressive efforts are impossible to flush away. Do you:

(a) stammer 'sorry' as you brush past her, vowing to avoid her for the rest of the evening?

(b) give her a wink and say nonchalantly, 'Oh, by the way, it's not mine?'

(c) stare her full in the face, as if to say that, as far as you're concerned, only wets pull the chain?

Allowing 3 points for (a) answers, 5 points for (b) and 8 points for (c), how did you score?

21-35 **You are an extremely nice, if somewhat boring person. Try to assert yourself a little more. At present you are what TBS experts refer to as a 'pathetic little wimp'.**

35-46 **Your toilet persona shows considerable potential. Keep it up and you'll make it to the executive toilet eventually.**

46-56 **You are exceptional. You are so self-confident that you probably never bothered with toilet-training at all. If you are not already a chat show host, the head of a nationalised industry or Prime Minister, you should be.**

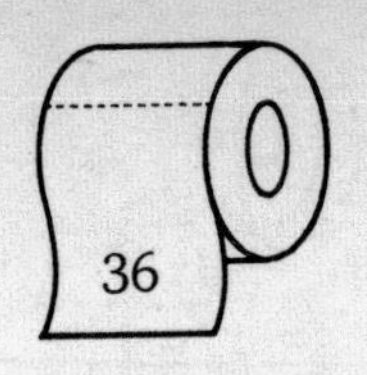

Letter from America

The following letter was received by a GI serving in Europe during the Second World War. It was from his mother, who lived in the mid-west of America.

'Pop has got a job, the first since you were born. What with that and your allowance, we decided to launch out a bit and build the bathroom we've always aimed for. It was finished a week ago. It's a swell job. In one corner is a large kind of tub, a bit bigger than the pig trough. That's for washing all over. On the same side as that is a smaller basin in which you can wash your face and hands. In the other corner is a pedestal arrangement for your feet. First you wash one in it and then you pull a chain and down comes a fresh supply of water for you to wash the other. And that's not all. The firm that sent out the fittings is generous. They sent us a mahogany frame, though they forgot the glass, and we've put this up in the parlour with the enlargement of granpop's picture in it. They also sent a solid board that makes a swell bread board and several rolls of writing paper . . .'

37

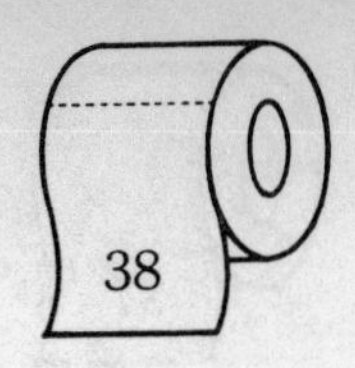

A name fit for a lavatory . . .

To give the flushing lavatory a name ringing with poetic resonance has tested the nation's greatest poets for over a century. Here are just some of the inspiring names that our great writers have worked on – but there is one odd one out in the list. See if you have the language-sensitivity to unmask the intruder.

The Sultan
The Lion Closet
Aeneas
The Closet of the Century
Deluge
Alerto
Niagara
Improved Infante
Original Burlington
Cascade
Tornado
Michelangelo
Subito
Corniche
Pluvius
Aquarius
Nerus
Sloane
Manresa
Caprice
Wentworth
Uniregal
Contour
Profile
Brasilia
Tulip
Tiara
The Rocket
The National
The Citizen
The Waterway
The Dolphin Patent Non-Splash Thunderbowl
The Westminster Portcullis
The Waterfall

The Corniche is of course a Rolls Royce motor car not a toilet – although the engine is reckoned to be as smooth running as the modern lavatory cistern.

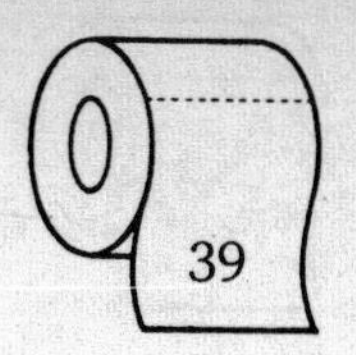

Early nursery rhyme

Little Robin Red Breast,
Sitting on a pole,
Niddle, Noddle,
Went his head,
And poop went his hole.

15p to spend a penny . . .

Such anachronistic headlines litter local newspapers reporting on affairs lavatorial, but for some it would be cheap at the price.

A school caretaker who visited a public convenience at Barnehurst, Kent, found there were no lights. While he was sitting on the lid-less pan, the whole thing fell to the ground and broke. He fell on top of it injuring his thigh on the broken pieces and was subsequently off work for five months. Eventually Mr Norman Edwards, who was still unable to dance, was awarded damages of £5,560 in court plus interest and legal costs. The bill for this particularly expensive pee went to Bromley Borough Council.

★ ★ ★

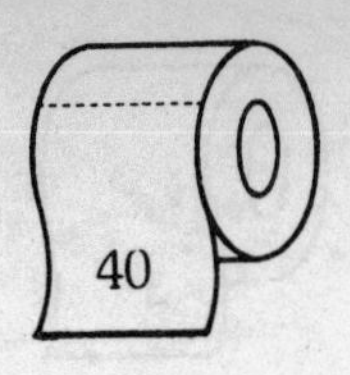

At Buxton, Derbyshire, spending a penny costs jittery L-drivers their examination fee and a chance of passing their test. The nearest loo to the motor test centre is half a mile away, and often they can't make it back in time for their test.

★ ★ ★

The seventy-three toilets and nine urinals run by Westminster City Council cost one and a half million pounds annually. Adding to the expense, that is, of course, borne by the tax payer, twenty people a year die in the toilets and the police are summoned to public lavatories ten times a day.

★ ★ ★

Road sweeper Michael Simpson was caught short on the job and discreetly went behind a garage to relieve himself. But he was reported by a woman who caught a glimpse of him, and the twenty-one year old was sacked on the spot from his job with Cleveland Council. After pleas by the public employees' union he was reinstated, but the lay-off cost him over two hundred pounds in wages. As he said, 'I'm pleased to have got my job back, but it's the costliest penny I'm ever likely to spend.'

★ ★ ★

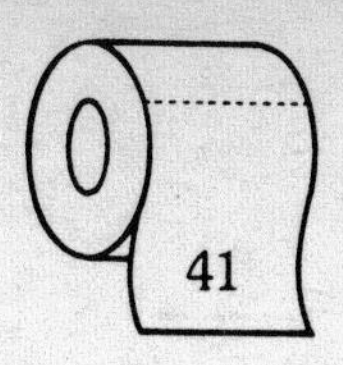

Little did a nightclub know that when Ed Shelton went into their toilet during the annual Christmas disco it was going to leave them over £100 out of pocket. When Ed sat on the toilet it collapsed and he fell to the floor, suffering concussion. He woke in the casualty ward of the local hospital, and even after he left he was still unable to sit down for two weeks. A court awarded him £127 to compensate for his concussion and lacerated buttocks.

For one unfortunate man a trip to the toilet cost him his freedom. When he went to a public lavatory in Brighton, he found himself padlocked in. Police and firemen worked with a will to free him but he was no sooner out than the police locked him up again. They'd recognised him as a man wanted on a theft charge, and that evening he was travelling under police escort to gaol in the North.

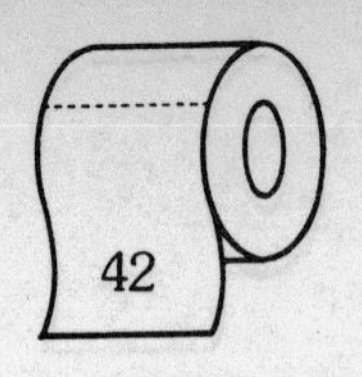

Loo and Non-Loo

Nothing reveals the character of a house and its occupants more infallibly than the smallest room. How stylish is your loo?

Loo

Cistern

The noiser and older it is the better.

Bowl

Shanks or Twyford pre 1950.

Paper

*White Andrex or newspaper (*Daily Telegraph*) squares. It is also rather amusing, if a little uncomfortable, to have old-fashioned Bronco.*

Paper dispenser

A plain wooden one that falls off when you try to use it.

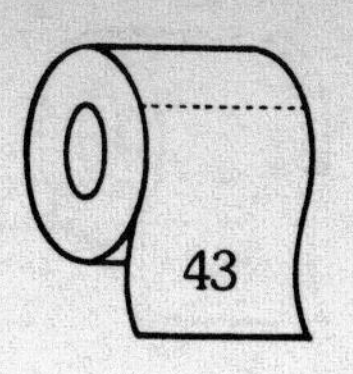

Non-Loo

Two-speed, silent job with blue water.

Any other bowls, particularly the dry German model.

Squares in boxes.

A musical loo roll.

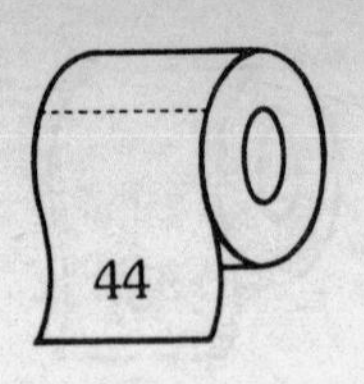

Loo

Hygiene

A few old smears around the bowl are generally regarded as rather stylish. If there's a basin, there should either be no soap at all or one caked in mud. A very old, dirty towel is acceptable.

Pictures

Hunting prints.
Original doodles by Hockney, Scarfe, Stubbs, Dufy, Princess Grace, Lionel Edwards or Noël Coward.
Colour photographs cut out of a gardening catalogue.

Photographs

Any snap by Cartier-Bresson.
Candid nude shots of celebrities (e.g. Jackie Onassis, Yoko Ono, the Duke of Edinburgh).

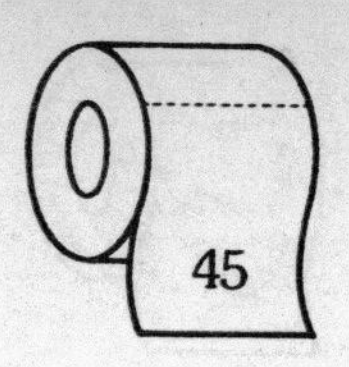

Non-Loo

Anti-smell devices.
Air expellers.
Nasty little bins for 'feminine hygiene'.
Hand dryers.
Fresh linen towels.

Hunting originals.
Silly toilet cartoons (e.g. 'Rodin's The Crapper' by Larry, 'The Man Who Asked For The Toilet In The Garrick' by H. M. Bateman, etc.).
Any original oil or watercolour, especially by the host or hostess.
Cute pictures of a little boy peeing or a dog biting a baby's bottom.
The talentless daubs of the child of the house.
An undiscovered masterpiece.

Old team photographs.
Heartwarming shots of the kids.
Anything by Bailey, Lichfield or Newton, the German pornographer.

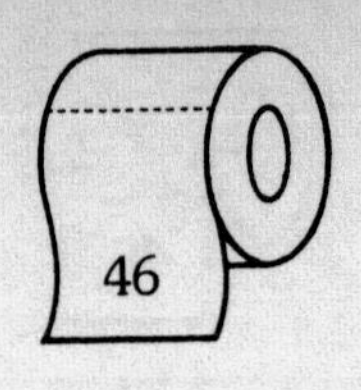

Loo

Objects

A loo brush.

Books

Up the Cistern
Giles Cartoons (but only pre 1963)
The Bible
Tristram Shandy
Blaise Pascal's *Pensées* (in French)
Wisden
Geoffrey Sale's *Hunter Chasers and Point-to-Pointers*
The Oxford Book of Aphorisms
Any books by Surtees
Diary of a Nobody

Magazines

Horse and Hound
The Eagle
The Sanitary Engineer
Illustrated London News
Variety
Evesham Journal

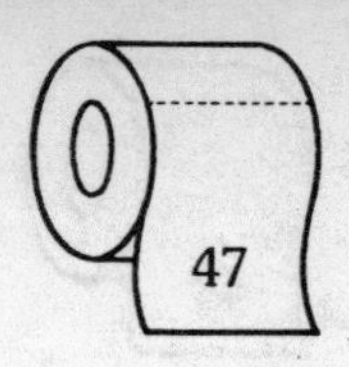

Non-Loo

Dolls that conceal a spare loo roll in their skirts.
Posters.
Any certificate or diploma.
A jumbo crossword.
A fake-fur, fur loo cover.
Calendars.
A pub bar mirror.

Stranks and Grant's *Are You Sitting Comfortably?*
The Sloane Ranger Diary
Any book published by *Private Eye* since 1975.
Snoopy cartoons
Any work by René Descartes
The One Minute Manager
Nigel Rees's *Graffiti Calendar*
Robert Lacey's *Princess*
Diary of a Somebody

Punch
Reader's Digest
Mayfair
Tatler

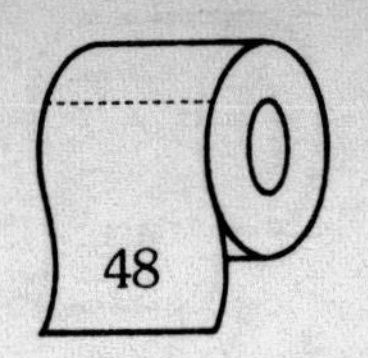
48

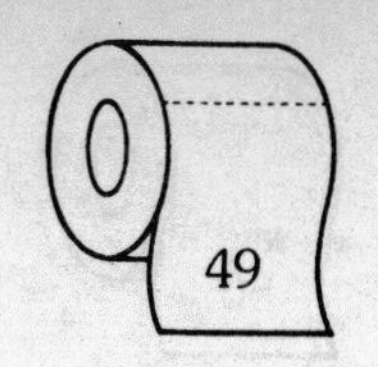

III

BLOW, WINDS

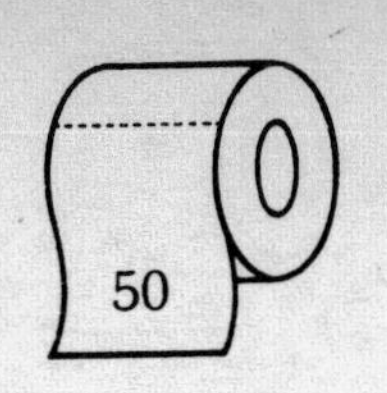
50

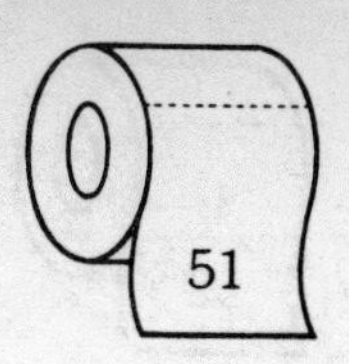

A flatulent plumber called Hart,
Could not get his blowlamp to start.
So he then struck a match,
Saying 'Now it'll catch' –
Thus extinguishing Hart, lamp and fart.

☆ ☆ ☆

There was a young girl of La Plata
Who was widely renowned as a farta.
Her deafening reports
At the Argentine sports
Made her much in demand as a starta.

☆ ☆ ☆

There was a young lady named Cager
Who, as a result of a wager,
Consented to fart
The whole oboe part
Of Mozart's Quartet in F Major.

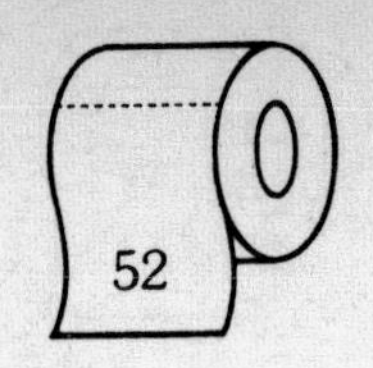

A Windy Etymology

Fart: *verb.* **Not in decent use. (Middle English *verten*, German *furzen*)**

1. (Intransitive) to break wind.
2. (Transitive) to send forth as wind from the anus.

One of the first recorded uses in *Cuckoo Song* (1250):

> 'Summer is ycomen in:
> Bulloc sterteth (leapeth)
> Bucke verteth.' (farteth)

Well-established by time of Chaucer:

> 'He was a little squeamish in the matter
> of farting, and satirical in chatter.'
>
> (of Absalon in *The Miller's Tale*)

And continuous usage from that time on:

> 'And to make sport I fart and snort.'
>
> (*pranks of Robin Goodfellow*, Attrib. Ben Jonson)

> 'The Earle of Oxford, making his low obeisance to Queen Elizabeth, happened to let a Fart, at which he was so abashed and ashamed that he went to Travell, 7 years. On his return the Queen welcomed him home, and sayd My Lord I had forgott the Fart.'
>
> (John Aubrey, *Brief Lives*)

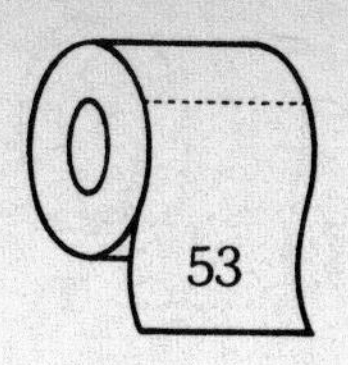

'He who farts in church
Sits on his own pew.'

(traditional schoolboy, still current)

Alternatives:

To break wind:

Early expression, brought back into current usage c. 1940.

'To break a little wind, sometimes one's life doth save,
For want of vent behind, some folke their ruin have!'

(Sir Thomas Moore)

'A man may break a word with you, sir, and words are but wind. Ay, and break it in your face, so he break it not behind.'

(William Shakespeare, *Comedy of Errors*)

To let one fly:

Middle ages and current.

'This Nicholas at once let fly a fart,
As loud as if it were a thunderclap.
He was near blinded by the blast poor chap.'

(Geoffrey Chaucer, *The Miller's Tale*)

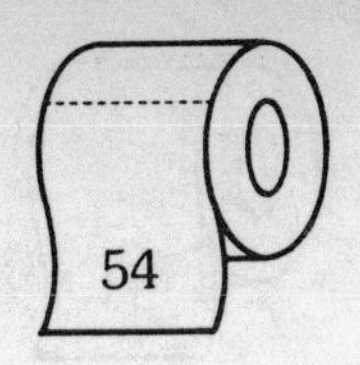

There's a smell of gunpowder:

British army euphemism from the nineteenth century. Draws analogy between firing a cannon and the breaking of wind in man. Possibly related to an earlier similar usage.

'Beware of thy hinder parts from gun-blasting.'
(*Book of Cutayse*, 14th century)

Make a trumpet of the arse:

Seventeenth century, fallen into disuse.

To backfire:

A modern euphemism that compares the noise of breaking wind with the backfiring of a motorcar.

SBD:

American teenage slang. Short for 'silent but deadly'. Origin obvious.

Master Robert is talking German:

Quoted by Sir Hugh Casson and Joyce Grenfell in *Nanny Says*. Racist euphemism, probably originating in First World War.

A breezer:

Common Australian noun. Comes from the original meaning of 'Breeze' – to blow softly.

A raspberry tart:

Noun. Rhyming slang dating from 1875.
Also *Bullock's heart* and *horse and cart*.

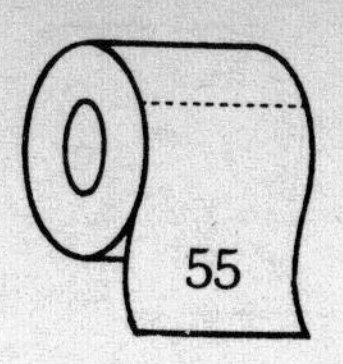

Shoot a bunny:

Origin unknown.

Bronx cheer:

Common American usage.

Boff:

North country slang. Derivation believed to be from 'to blow off'.

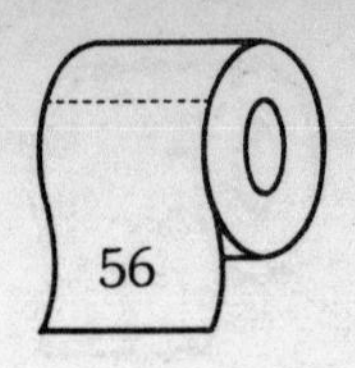

Some Strange farting facts

The great Greek scholar Dr Gilbert Murray always translated the verb 'to break wind' as to 'blow one's nose.'

Prostitutes in fifteenth-century France, known as *Filles publiques*, were allowed to pay the toll for crossing a bridge either with money or by letting fly a fart.

Honoré de Balzac, when pondering his highest ambition, wrote: *'I should like one of these days to be so well known, so popular, so celebrated, so famous, that it would permit me to break wind in society and society would think it a most natural thing.'*

Saint Augustine is said to have had such strong control of his backside that he could break wind at will, and could follow the tone of chanted verses.

In the Middle Ages in Northumberland, there was a competitition known as 'Farting for the Pig'.

The Emperor Claudius is reported by Suetonius to have seriously considered passing a decree that man could pass wind at table after the Emperor was told the story of a person whose modest retention cost him his life.

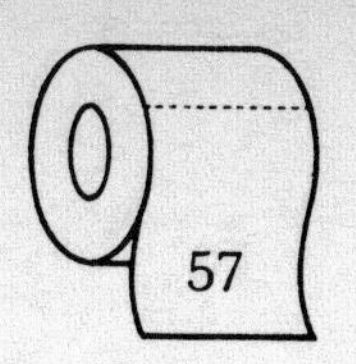

PETOLOGY – *An ancient art rediscovered*

Like astrology, palmistry, the Tarot and I Ching, the art of petology was practised centuries ago and has only recently been seen as having any significance whatever today.

Petology – literally the science and study of fart-reading – was first reported in what is now the Flemish part of Belgium during the days of the Roman empire. Among the lowlanders, the village petologer was revered in much the same way as the local doctor, schoolmaster or parson would be.

Enrik Breeghel, the so-called 'Bard of Feynoorde', dedicated an entire poem to petology, 'Le Grand Pétologue des Pays Flamands', and many believe that it was on the advice of the court petologer at The Hague that King Froeke of the Lowlands declared war against the Prussians in 1206. Later, the ill-fated Flemo-Prussian conflict was dubbed 'The Twelve Hour War'.

Although it was claimed that the Flemish petologist had misread the winds, the catastrophe served to discredit the art of petology until relatively recently.

The revival of interest in the ancient art has been spearheaded by a group in California who, during the late sixties, lived exclusively off cabbage and lentils and claimed that they had discovered totally new vibrations within themselves. Yet, despite the support of international figures such as Pierre Trudeau, George Harrison, and President Zia of Pakistan, the campaign to get petology accepted as a *bona fide* way of telling character and the future has been an uphill one.

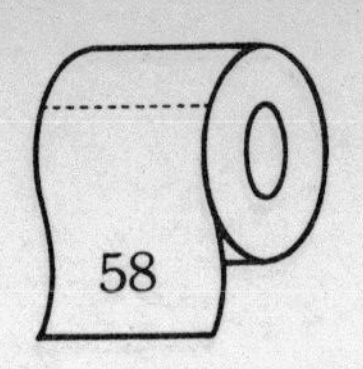

'People still treat us like we've just done a Silent Enemy or even a Cruise,' a spokesman said, 'but all we want is a chance to prove that petology can help cure the problems of modern life.'

For non-petologists, here is a summary of how your own farts can speak volumes about your character, together with a list of the type of people most often associated with them.

The Silent Enemy

You are devastatingly effective, even if it means offending people, as you get around. You may not be popular – particularly in lifts or crowded railway carriages – but there's no ignoring you.

(Civil servants, waiters, hairdressers, newsreaders, booksellers, cabinet ministers, literary agents, members of any theatrical audience.)

The Machine Gun

You're full of vim and bounce and you're very difficult to stop once you've got going. You can be great fun at parties and you can be a show-stopper in a crowd.

(Secretaries, TV comedians, sales reps, joggers, anyone on the production line of a factory, housewives, disco dancers.)

The Cruise

A new petological term used to describe the type of person who keeps moving around a room, laying farts as he or she goes, making it more difficult to aim a counterstrike. You are not a popular person, particularly when it is discovered that you only

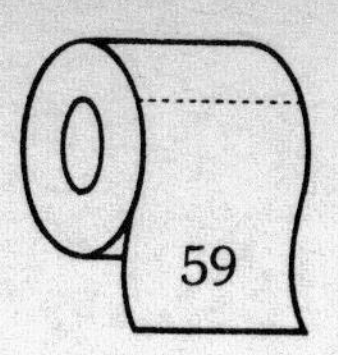

behave like this when you're in other people's houses, never your own.

(Traffic wardens, bank managers, doctors, bus conductors, plumbers, butlers, and almost all back-bench MPs.)

The Air Raid Siren

The long, high-pitched complaining whine that is your hallmark is a signal for all right-minded people to clear the room. If only you had the self-confidence to come right out with it or the decency to keep it in.

(Gossip columnists, middle management executives, writers, groupies, actors, VAT inspectors, teachers.)

The Depth Charge

You have awe-inspiring confidence. Your mere arrival on the scene has an immediate effect on all around you, who are either amused by your cheerful, extrovert behaviour, or are fearfully jealous of your nerve and dash.

(Prostitutes, taxi-drivers, actresses, members of the Royal Family, stand-up comics, rock stars, barmen and barmaids.)

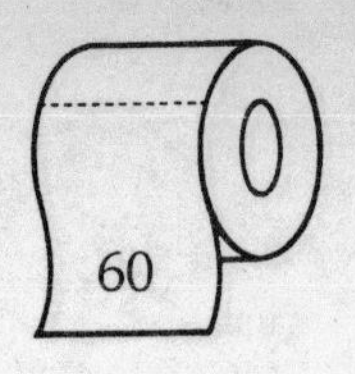

Le Petomane

In the 1890s Le Petomane took Paris by storm. His earnings at the Moulin Rouge were double those of Sarah Bernhardt and treble those of lesser stars. The particular show business genre to which he subscribed was that of a farting performance on stage. He was a confident artiste and told his children:

'I never had stage fright before going on – not even on my opening night at the Moulin Rouge.' The following is an extract from his son's book about his famous father.

'This is how Le Petomane presented himself with an ease and good humour which worked beautifully on the public.

' "Ladies and gentlemen, I have the honour to present a session of Petomanie. The word Petomanie means someone who can break wind at will, but don't let your nose worry you. My parents ruined themselves scenting my rectum."

'During the initial silence my father coolly began a series of small farts, naming each one. "This is a little girl, this the mother-in-law, this the bride on her wedding night (very little) and the morning after (very loud), this the mason (dry – no cement), this the dressmaker tearing two yards of calico (this one lasted at least ten seconds and imitated to perfection the sound of material being torn), then a cannon (Gunners stand by your guns! Ready! – Fire!), the noise of thunder." Etc., etc.

'Then my father would disappear for a moment behind the scenes to insert the end of a rubber tube, such as are used for enemas. It was about a yard long and he would take the other end in his fingers and in it place a cigarette which he lit. He would then smoke the cigarette as if it were in his mouth, the contraction of his muscles causing the cigarette to be drawn in and then the smoke blown out. Finally my father removed the cigarette and blew out the smoke he had taken in. He then placed a little flute with six stops in the end of the tube and played one or two little tunes such as Le Roi Dagobert, *and, of course,* Au Clair de la Lune. *To end the act he removed the flute and then blew out several gas jets in the footlights with some force. Then he invited the audience to sing in chorus with him.*

'From the beginning of the "Audition" mad laughter had come. This soon built up into general applause. The public and especially women fell about laughing. They would cry from laughing. Many fainted and fell down and had to be resuscitated.'

From Jean Nohain and F. Caradec, Le Petomane, *1967.*
Trans. Warren Tute, Souvenir Press.

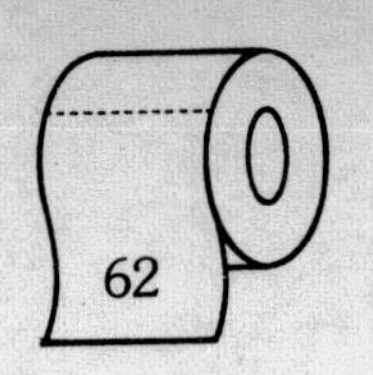

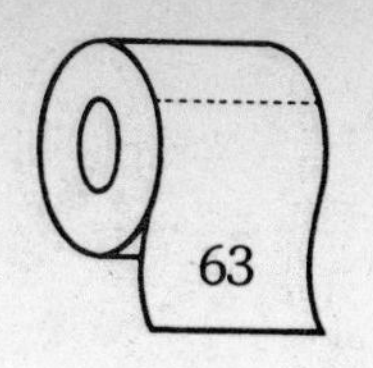

IV

ROMANTIC MOTIONS

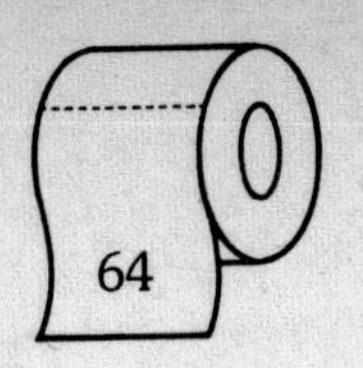
64

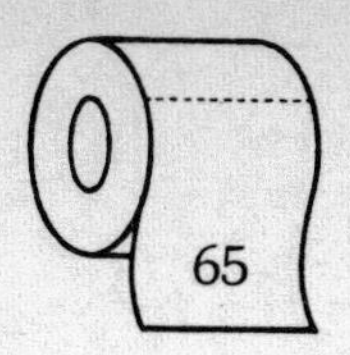

A plumber from Lowater Creek
Was called in by a dame with a leak;
She looked so becoming
He fixed all her plumbing
And didn't emerge for a week.

☆ ☆ ☆

There was an old fellow from Kilbride
Who fell down a privy and died.
He had a younger brother
Who fell down another
And now they're interred side by side.

☆ ☆ ☆

There was a young student of Queens
Who haunted the public latrines.
He was heard in the john
Saying 'Bring me a don,
But spare me those dreary old deans.'

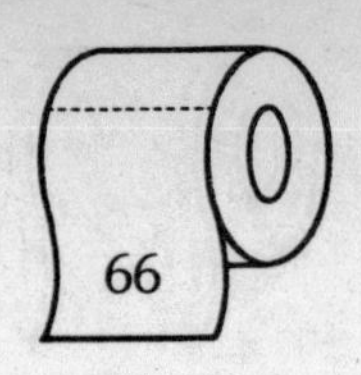

Seats of Love

While the smallest room is not always associated with matters of love, occasional tendernesses have been known to take place there.

The first time James Joyce saw Marthe Fleischmann, the woman who was to become his lover, was in 1919. 'She was in a small but well-lit room,' he was later to write, 'in the act of pulling the chain.'

'Only a few hours after I had been introduced to the House (of Commons), when I was still wandering about in a daze, and lost, Chips (Channon) kindly showed me round the most important rooms – the Members' lavatories. This was an act of pure, disinterested, sisterly friendship, for we had no physical attraction for each other.'

Tom Driberg, *Ruling Passions*, 1977.

One of the most spectacularly randy men in history, Cesare Borgia spent most of his wedding night on the lavatory. As a joke, one of the guests had substituted some laxatives for other pills that Borgia had ordered from the apothecary.

One of the most legendary loo love scenes of recent times was the highlight of the soft porn epic Emmanuelle. *Taken short during a flight on an airliner, Emmanuelle takes a male passenger with whom she has struck up an instant rapport to the loo at the back of the aircraft. There they both join the mile-high club with appropriately ecstatic sound effects.*

Poetry in Motions

Oh Cloacina Goddess of this place
Look on thy servant with a smiling face
Soft and cohesive let my offering flow
Not rudely swift nor obstinately slow.

Poem sometimes found in Victorian lavatories

★ ★ ★

Use me well, and keep me clean,
And I'll not tell what I have seen.

Words on the bottom of an early nineteenth-century jerry-can

★ ★ ★

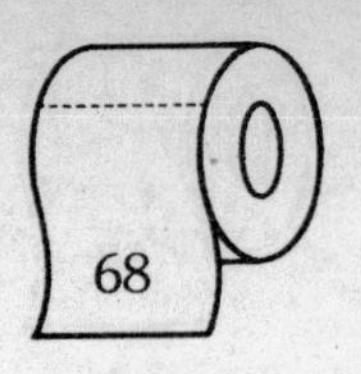

When you would have your urine pass away,
Stand not upright before the eye of day;
And scatter not your water as you go
Nor let it, when you're naked, from you flow:
In either case 'tis an unseemly sight:
The gods observe alike by day and night:
The man whom we devout and wise may call
Sits in that act, or streams against the wall.

Thomas Cooke's translation of Hesiod's *Works and Days*

I ran upon life unknowing, without either science or art,
I found the first pretty maiden but she was a harlot at heart,
I wandered about the woodland, after the melting snow,
'Here is the first pretty snowdrop' – and it was the dung of a crow.

Alfred Tennyson

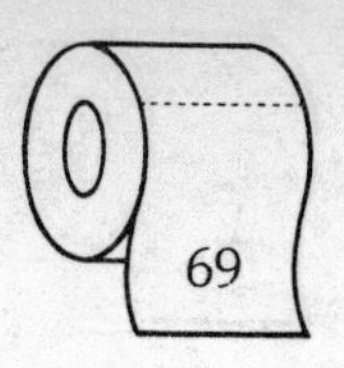

Stool Pigeon

During the last world war, Harold Harris, a captain in the Intelligence Corps, received reports of a spy working from the Isle of Lewis in the north of Scotland. Every night at 9.30 p.m. the spy flashed his light out to sea, presumably to a lurking submarine which recorded his message. The signal consisted of a series of flashes, followed by a long continuous beam, followed by another series of flashes. About 9.40 each evening the message was complete.

Harris reported the incident to his superior, Major Solvesen, the Security Officer. The area from which the signal came was so remote that the local constable had great difficulty getting there, and, after due consideration, the two officers decided to report to the War Office that, although they had seen the flashing signal, they had been unable to trace it to its source and they did not deem the matter worthy of further investigation. In fact, in his six page foolscap report Solvesen went so far as to say that he felt in the remote area from which the spy operated there could be no important information for him to pass on, and if he was indeed a spy it would be best to leave him there where he could do much less damage than anywhere else in the British Isles, and where he was occupying the time of one German submarine.

The War Office felt differently. A Mr Price was smartly despatched from London to investigate. From his city desk he had not quite appreciated the remoteness of the spy's base, but undeterred by the discovery, he set off to hunt him down, accompanied by a lance corporal disguised in sports jacket and flannels. Arriving after a day or two's travelling by land and sea in the supposed area of enemy operations, Price visited the

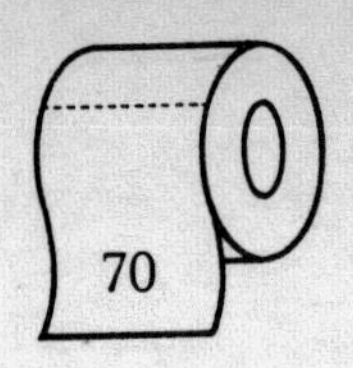

points from which the spy's light could be seen, set up highly sophisticated direction-finding apparatus, and instructed the lance corporal on how to work it. Bearings were plotted, and the light actually pin-pointed on the map. On the following day the chase reached its climax. With some difficulty, because of the extreme remoteness of the region, the spy was tracked down.

An ancient crofter was the supposed ally of German might. A man of regular habits, he had no indoor toilet in the house, and every night, before he went to bed, he walked some twenty yards across his land to a kind of cave, swinging his lantern in time with his stride. He then put the lamp on a ledge, while he squatted down, and swung it again on his return. His story was corroborated by the evidence in the cave.

The War Office reckoned it was a satisfactory conclusion to a difficult case, even if it had been a long journey (about 1,600 miles there and back) to discover a crofter's lavatory.

Originally told by Harold Harris in *The War on Land*, edited by Ronald Lewin, Hutchinson, 1969.

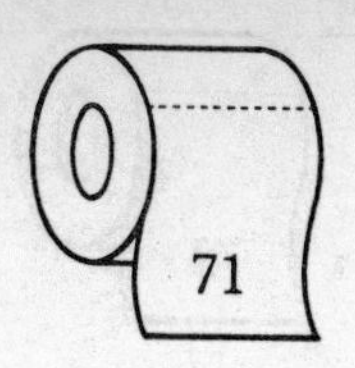

Small Room Superstitions

In America during the nineteenth century it was said to be good luck for a man to make the shape of a cross while urinating on the ground.

☆ ☆ ☆

If a man allows his pee to fall on his shadow, ill fortune will occur.

☆ ☆ ☆

The German seventeenth-century writer Johannes Frommann had this useful advice for those of his readers who had excrement left on their doorstep by a neighbour: 'If fried beans be thrown into excrement, for each bean thus wasted a pustule will appear on the fundament of the thrower.'

☆ ☆ ☆

In mediaeval England it was believed that 'a woman who urinates upon nettles will be peevish for a whole day'.

☆ ☆ ☆

A familiar European superstition was that a pregnant woman could tell whether she was going to have a boy or a girl by first urinating on a seed of wheat and then on a seed of barley. If the wheat sprouted first, she was to have a boy; if the barley, a girl.

☆ ☆ ☆

During the Middle Ages, it was believed that Lapland witches could hold a ship in their power unless it had previously been protected by a liberal smearing along the seams on the inside of the ship with 'the ordure of virgins.'

☆ ☆ ☆

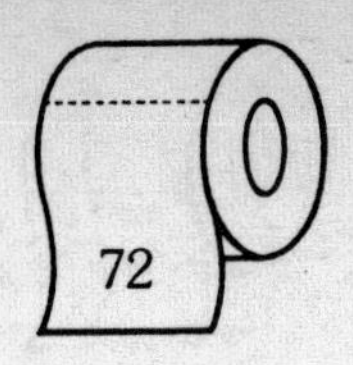

The thigh bone of a man who has died a violent death will, if emptied of its marrow, filled with human ordure, sealed up with wax and placed in boiling water, cast a spell on the man whose ordure it is. He will be forced to evacuate just as long as the bone is kept in boiling water. In this way, an enemy can be made to mess his bed every night, if required.

There is an ancient superstitition that if a robber 'eases himself' at the place where he is about to commit a robbery, then he will be secure from interruption for some time.

During the Middle Ages, a person interrupted while 'easing himself' near a highway or footpath was obliged, if the passer-by said 'reverence', to take off his hat with his teeth and, without moving his position, throw it over his head where, not surprisingly, it would normally land messily. This was considered a punishment for a breach of delicacy in choosing a public place. Anyone refusing to obey the custom would be pushed backwards with unpleasant results.

Traditional wisdom in France says that a dream of ordure means that good fortune is on the way.

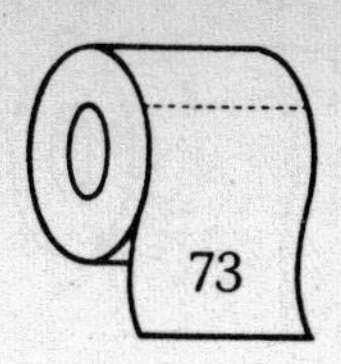

It is said that, until the seventeenth century, if you cried out 'in pain!' three times on a public vehicle and it failed to stop, you were not obliged to find a lavatory but could do it where you were.

☆ ☆ ☆

When cursed by a witch's spell, it is advisable to urinate through a wedding ring.

☆ ☆ ☆

The Plumber's Tale: A Eurofable

There was a young plumber of Leigh
Who was plumbing his girl by the sea.
She said, 'Stop your plumbing,
There's somebody coming.'
'I know,' said the plumber. 'It's me.'

★ ★ ★

Il avait un plombier, Francois,
Qui plombait sa femme dans le bois,
Dit-elle 'Arretez!
J'entends quelqu'un venait.'
Dit le plombier, en plombant, 'C'est moi.'

★ ★ ★

Es gibt ein Arbeiter von Tinz,
Er schaft mit ein Madel von Linz.
Sie sagt, 'Halt sein plummen,
Ich hore Mann kommen.'
'Jacht, jacht,' sagt der Plummber. 'Ich binz.'

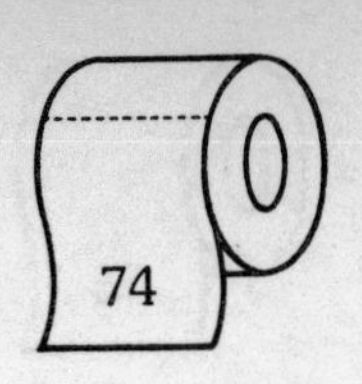

The Naming of Parts

Any expert on the small room should know the words associated with it. Try the following word quiz:

1. *Coprolite* is
   (a) The boy servant of a Greek nobleman
   b) A fossilised turd
   c) A base form of copper

2. *Whigget* is
   a) A lock of pubic hair
   b) An impolite written reply
   c) The dung to be found on a dog's bum

3. *Urcolate* is
   a) Pitcher-shaped
   b) The colour of urine
   c) A form of coffee preparation

4. *Vagitus* is
   a) A type of cabbage
   b) The irritation caused by cystitis
   c) The noise a new-born baby makes

5. *Fumet* is
   a) Deer dung
   b) The smoking compartment on a French train
   c) Lint

6. *Ted* is

a) An incompletely castrated bull
b) To spread manure
c) A young antelope

7. *Renifleur* is

a) One who gets sexual pleasure from body odours
b) A homosexual prostitute
c) The generic name for water plants

8. *Uranophobia* is

a) Fear of invasion from outer space
b) Fear of urine
c) Fear of heaven

9. *Carminative* is

a) Fattening
b) Slimming
c) Gas inducing

10. *Merkin* is

a) A woman's pubic wig
b) A pig's dung
c) A Jacobean garment

11. *Zippertrauma* is

a) A psychological term for a morbid fear of zippers
b) A medical term for damage to the penis caused by a zipper
c) Hollywood slang for someone who's 'in a muddle'

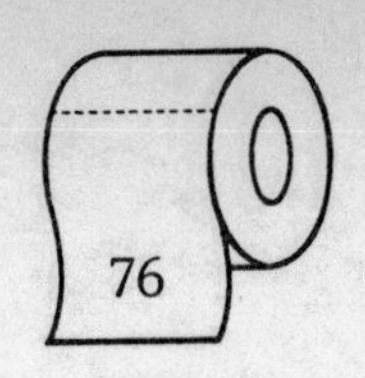

12. *Urning* is
   a) The purification of milk
   b) A type of early lavatory bowl
   c) A male homosexual

13. *Trocar* is
   a) A tool inserted into cattle to relieve gas
   b) A eunuch who can still make love
   c) A type of wind instrument

14. *Pissasphalt* is
   a) Semi-liquid bitumen
   b) A seventeenth-century cure for the clap
   c) A mixture of wine and asphalt

15. *Furbelow* is
   a) The urine of a young labrador
   b) A page boy
   c) Ruffs around a woman's dress

16. *Urostyle* is
   a) A particularly long last vertebra
   b) A chemical testing device for urine
   c) A Belgian term for modern fashion

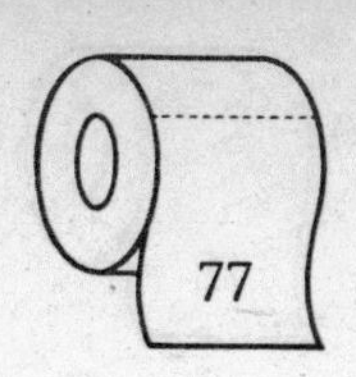

17. *Steatopygic* is
   a) Aroused by fat buttocks
   b) The shape fat buttocks make in sand or in a mould
   c) Possessing fat buttocks

18. *Rectalgia* is
   a) A yearning for lost purity
   b) A pain in the bottom
   c) Heartburn

19. *Farctate* is
   a) Full or stuffed
   b) Suffering from wind
   c) To digest

20. *Toilinette* is
   a) An Edwardian public lavatory attendant
   b) Woollen cloth used for waistcoats
   c) A cheap, open public toilet for women

1. b); 2. c); 3. a); 4. c); 5. a); 6. b); 7. a); 8. c); 9. c); 10. a); 11. b); 12. c); 13. a); 14. a); 15. c); 16. a); 17. c); 18. b); 19. a); 20. b)

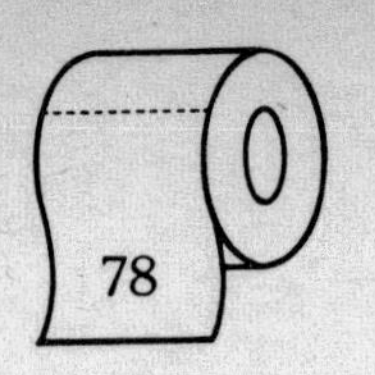
78

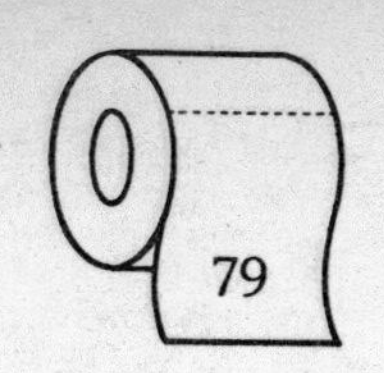

V

SITTING ON THEIR DIGNITY

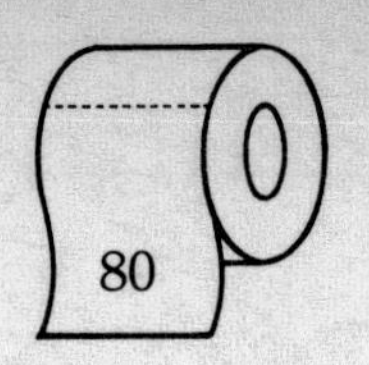
80

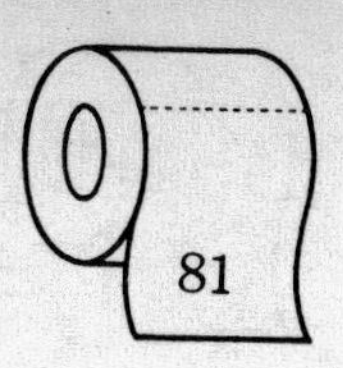

The Duchess when pouring the tea,
Once asked 'Do you fart when you pee?'
I replied with some wit,
'Do you belch when you shit?'
And I think that was one up to me.

There was a young fellow called Chivy,
Who, whenever he went to the privy,
First solaced his mind,
Then wiped his behind,
With some well chosen pages of Livy.

''Tis my custom', said dear Lady Norris,
'To beg lifts from drivers of lorries.
When they get out to piss,
I see things I miss
At the wheel of my two-seater Morris.'

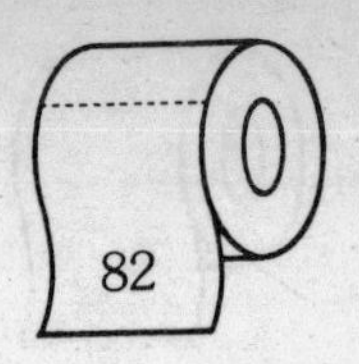

Royal Flush

It is a sadly inescapable fact that even royalty has to go to the lavatory sometimes. In fact, the royal toilet – or 'garderobe' as it's called in the best circles – has played a significant part in our history. Now the story can be told . . .

King Edmund Ironside, according to the historian, Guydo, was engaged on strictly non-royal business when he met his end. We are told that the murderer 'awaytynge his time, espyed when the king was at the withdraught to purge nature, and with a spere strake hym into the foundement, and so into the body'.

King Edmund was not the last king to die at his 'stool of easement'. Both James I of Scotland and George II continued the tradition.

Richard III was in the royal loo when he was advised by a page that a man called Tyrell would be prepared to murder the princes in the tower on the King's behalf. 'Upon his words King Richard arose, for this communication had he sitting at the draught, a convenient carpet for such a counsell.'

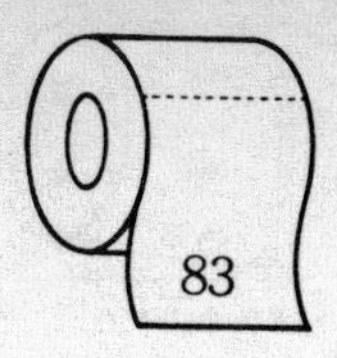

King James V of Scotland set great store by the comfort and luxuriousness of the royal 'close-stool'. Whereas his predecessor spent a mere 8d (eight old pennies) for the construction of a stool of ease in 1501, James V spent the princely sum of £52.2.6d for a pavilion covered in green damask.

☆ ☆ ☆

Henry VIII was similarly expansive. In 1547, he ordered from Mr Grene, the royal coffer-maker, a garde-stool of suitable style and dimensions. When completed, it was covered with black velvet and garnished with ribbons and fringes. The seat elbows were covered with white 'fuschan' filled with down and no fewer than 2,000 gilt nails were used in the construction. The King sat and was pleased.

☆ ☆ ☆

Charles II and his court were less interested in the royal toilet. According to the diary of Anthony à Wood, the King's court made quite an impression when they spent the summer of 1665 in Oxford in order to avoid the plague.

'Though they were neat and gay in their apparrell,' he wrote, 'yet they were very nasty and beastly, leaving at their departure their excrements in every corner, in chimneys, studies, colehouses, and cellars.

Queen Victoria and Albert, the Prince Consort, were most concerned about the sanitary conditions at Windsor Castle. Hearing that, in 1844, no fewer than fifty-three overflowing cesspools were found at Windsor Castle, a programme to replace the old Hanoverian commodes with up-to-date water closets was instigated. Some are still in action at the castle.

The Queen was also, she said, concerned that the people who worked on the royal estates were insufficiently catered for in this area, and lavatories were built at strategic points around the grounds. No one dared point out that the new lavatories were also convenient for the Queen herself who , as she got older, was finding it difficult to take a constitutional without strategic resting points.

Edward VII, while still the Prince of Wales, was seriously ill from typhoid on one occasion. On recovering, he became obsessively interested in toilet matters. 'If I were not a prince, I would be a plumber,' he said on one occasion.

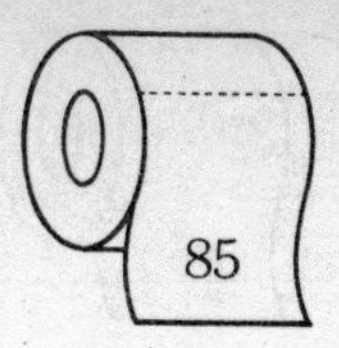

Edward VIII inaugurated an important change within the royal household. The person in charge of the royal toilet who, in the days when he catered for the personal needs of the monarch, was called the Yeoman of the Stole Chamber, then, as the job description changed with Charles II, became known as the Master of the Great Wardrobe, was now to be called the Usher of the Robes. George VI, for reasons best known to himself, changed this to Groom of the Robes and this important position is still occupied by a key member of the royal staff.

Queen Elizabeth II has, like her illustrious forbears, taken an interest in these personal matters, but no effort or expense is spared to protect the Royal Family from evidence that their subjects have private needs as well. Whenever the Queen goes on tour, all signs pointing to public conveniences are painted over. (Some royal observers believe that many of the people lining the route, jumping up and down and craning their necks, are merely trying to find the carefully concealed public lavatories.)

At one town hall that the Queen was to visit, it was discovered at the last moment that, even with the doors closed, the sound of flushing loos in the gents could distinctly be heard in the main hall where the Queen was to be received. A silencer was hurriedly installed.

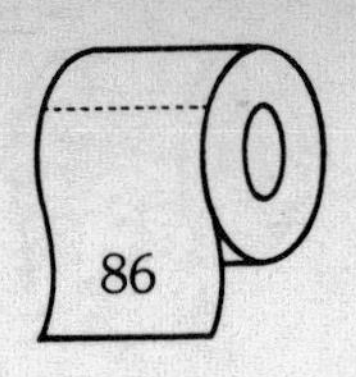

The Duke of Edinburgh has been typically forthright on the question of toilets. On one occasion, he said, 'This is the biggest waste of water by far. You spend half a pint and flush two gallons.'

On the other hand, a certain amount of wastage is in evidence on royal tours. When the Duke visited an airshow in Buckinghamshire, the local organisers arranged for an expensive hospital de-ioniser to be installed so that the royal guest's visit would be entirely germ-free.

And when the Queen and the Duke of Edinburgh visited the Commonwealth Games in Canada, they enjoyed the comforts of a custom-built loo. The organisers of the games had been horrified to discover that the nearest toilet to the royal box was over seventy yards away. To avoid frequent, unscheduled royal walkabouts, which would be seen by curious viewers all over the world, a new loo was installed, complete with thick carpeting, subdued lighting and up-to-date issues of *Country Life*. A slightly unfortunate side-effect of the royal loo was that its construction required part of the stadium to be underpinned with concrete.

The final cost of the royal wee was said to be £23,000.

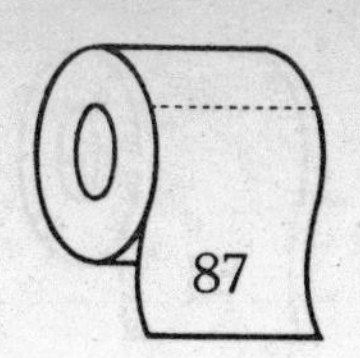

Programme for Official Opening of the Abakrampa Public Latrines

1. Public procession through the streets to the latrine led by Abakrampa Brass Band.
2. Guests to take seats.
3. Opening prayers by Reverend Abonyi (Methodist Church).
4. Introduction of Chairman and other important personalities by Mr S. G. Gyandoh, Snr.
5. Music.
6. A short history behind the project by the General Secretary.
7. Address by Administrative Office (Central Region).
8. The uses of this type of latrine, by Mr Ocloo.
9. Music.
10. The sod to be cut by a representative of the NLC and inspection of the latrine by the guests.
11. Omanhen's address.
12. Vote of thanks by the Clerk of the Council.
13. Music.
14. Light refreshment at the Community Centre.

Quoted in *World Medicine*

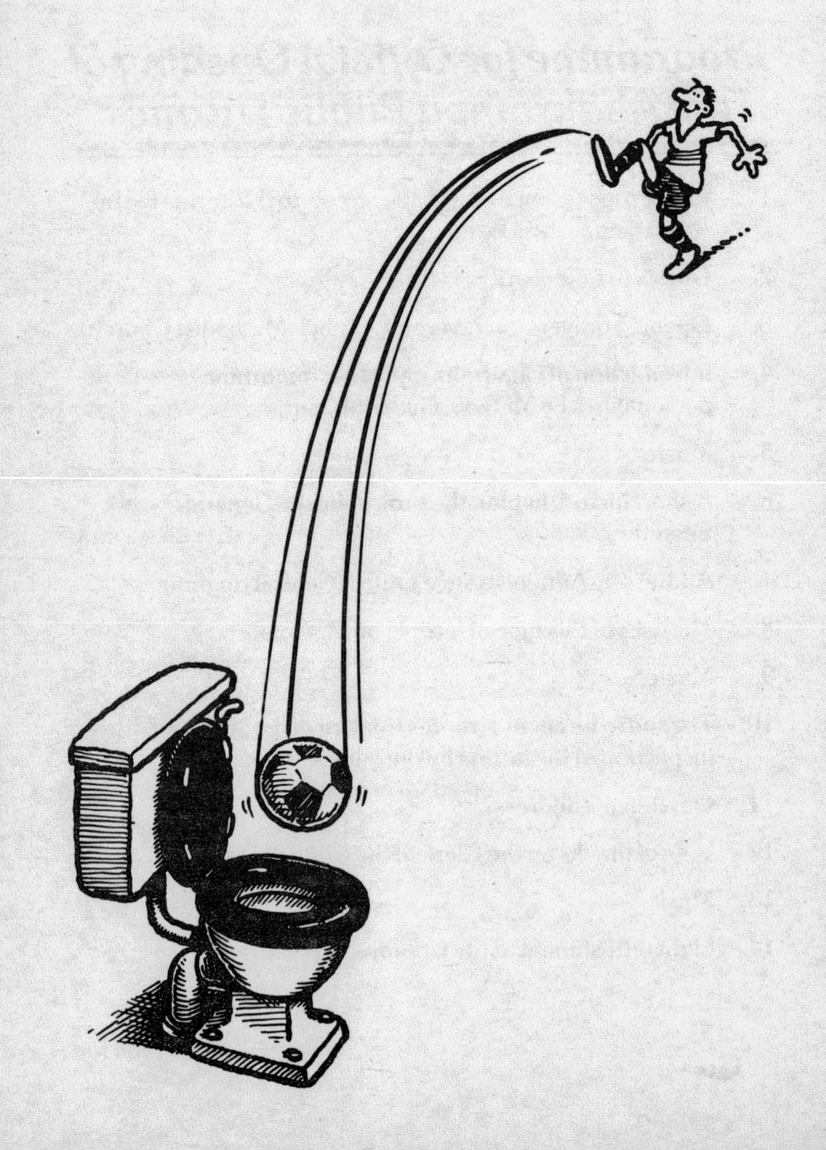

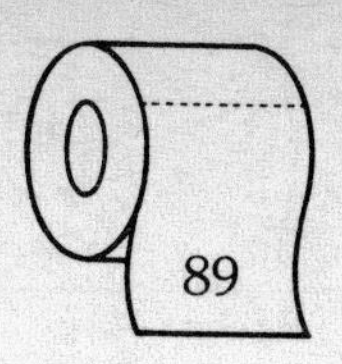

From the Council Chamber . . .

Indicating how thoroughly they did their job, he told the Council that at their last meeting, all the members were engaged, for a time, on counting sheets of toilet paper to see which firm offered the best value.

Liverpool Echo

A sub-committee is to consider the question of alterations at the village hall so that toilets can be used for football matches.

Daily Mirror

'When I was living in a house with an upstairs toilet, I applied for a grant to construct a downstairs toilet, and the social work department wrote back to say, despite my wheelchair, I didn't qualify but would I like a knitting machine instead?'

Edinburgh Evening News

Councillor A. J. Robson told Monday's meeting of the rural council, 'Today was the first day for nine days that some residents have been able to use their toilets. Something has got to be done to relieve this situation.'

Folkestone Herald

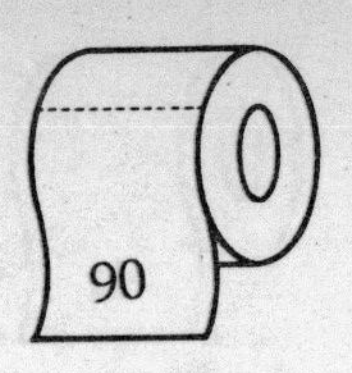

A trip down the sewers of history

Waste disposal in London

The stink of London for all but the last two decades of the great city's history would have been abhorrent to the sensitive noses of today. The earliest form of waste disposal was simply to empty the pot out of the window. Edward III observed that when riding up the Thames he 'beheld dung and lay stalks and other filth accumulated in divers places and noticed fumes and other abominable stenches arising therefrom'.

No serious attempt was made to tackle the problem until the reign of Richard II. He ordained in an act of 1388 that all dung and filth should be carried away to where it was not a nuisance and no one was to empty piss out of the windows into the streets. Swine that filthied the streets were to be killed on sight (c.f. contemporary treatment of dogs). But despite introducing the nicety of the handkerchief to these shores. Richard proved unequal both to the task of keeping his crown and to that of clearing up the streets of London.

In the annals of sewer history, the Great Fire of 1666 was a great opportunity missed. Despite the replanning of the city,the new sewers were quite unequal to the strain on the system imposed by the ever-growing population, and conditions remained as bad as previously, or worse, until the beginning of the nineteenth century, when there were signs of a wind of change. One of the indications can be seen in the following strange course of events: up until 1815 the discharge of offensive matter into the sewers was a *penal offence*; it then became permissible to drain houses into the sewers and in 1847 it became an offence *not* to drain a house into a sewer.

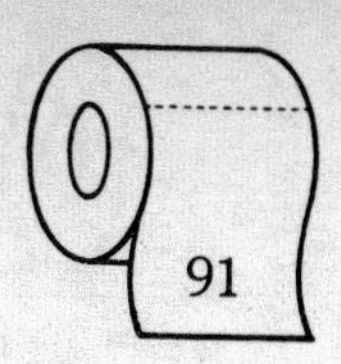

Matters came to a head in 1858, known as the Year of the Great Stink. Owing to the fast-growing population of the metropolis, the state of the air was worse than ever. The stench from the Thames in the House of Commons was so bad that sheets soaked in disinfectant had to be hung between the House and the river, and the committee rooms on the riverside had to be abandoned as unusable. And it wasn't only Thames dwellers who suffered: Hackney Brook, once a pleasant enough stream, was now no less than an open sewer. Gladstone and Disraeli, usually arch-rivals, agreed on this one issue: something must be done about the great stink.

The saviour of the city proved to be one Sir Joseph Bazalgette. Bazalgette was working on plans for five great sewers: three intercepting sewers north of the river, and two in the south – 'in total, a hundred miles of sewers running from east to west. The Metropolitan Board of Works was set up to build on these foundations and by 1919 the city could be judged to have an adequate system of sewers. It is these sewers that form the basis of the system that we take for granted today.

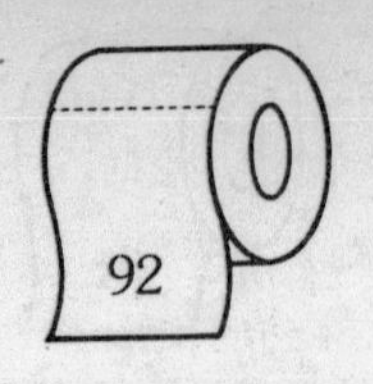

True or False?

When the final episode of *Mash* was shown on American television there was a record set for toilet use.

(True. An emergency 700,000 extra gallons had to be pumped into the system to cope with the enormous, simultaneous flush of all those who'd been holding out until the last second of the programme.)

Einstein claimed that his theory of relativity came to him while he was astride the lavatory.

(False. Although there are those who claim that another great inventor, Sir Isaac Newton, developed his theories on the lavatory and that the famous story about apples is a euphemism.)

'*Tant pis, tant mieux*' is a common French phrase meaning 'if my aunt pees she will feel better'.

(False. She won't feel better at all. Aunts frequently suffer from cystitis.)

Byron was banned from Long's Hotel in Bond Street for pissing in the hall.

(True. There is no evidence that it was his dog Bosun as one scholar claimed. The dog had already been dead a number of years before this incident.)

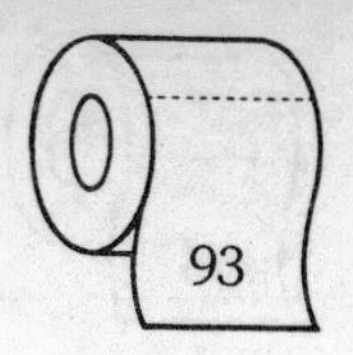

Straining for Effect

Some public quotes on private places.

'I once asked Tony Armstrong-Jones how he got on with all those personalities and he said: "My dear boy, I simply think of them sitting on the loo: works wonders." '

Terence Donovan

★ ★ ★

'After dinner with Sir W. Pen, my wife and Mary Batelier to the Duke of York's house, and there saw *Heraclius* which is a good play . . . My wife was ill and so I was forced to go out of the house with her to Lincoln's Inn Walks, and there in a corner she did her business, and was by and by well, and so into the house again.'

Samuel Pepys

★ ★ ★

'Life is like a sewer. What you get out of it depends on what you put into it.'

Tom Lehrer

★ ★ ★

'Never lose an opportunity for pumpship.'

Duke of Wellington

★ ★ ★

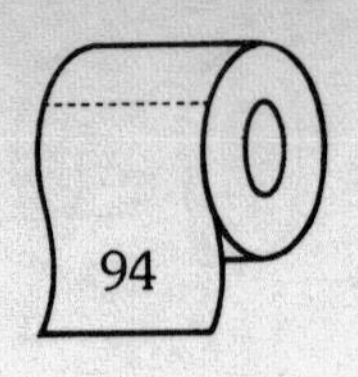

'Not so much goodbye as *au revoir*.'
Noël Coward (on emerging from the loo of a yacht lying in particularly smelly water).

★ ★ ★

'Quilted seats retain a bad smell. No sir, there is nothing so good as the plain board.'
Dr Johnson

★ ★ ★

'I knew a gentleman who was so good a manager of his time, that he would not even lose that small portion of it which the call of nature obliged him to pass in the necessary-house; but gradually went through all the Latin poets, in those moments. He bought, for example, a common edition of Horace, of which he tore off gradually a couple of pages, carried them with him to that necessary place, read them first, and then sent them down as a sacrifice to Cloacina: there was so much time fairly gained: and I recommend you to follow his example. It is better than doing what you cannot help doing at those moments.'
Lord Chesterfield to his son, 1747

★ ★ ★

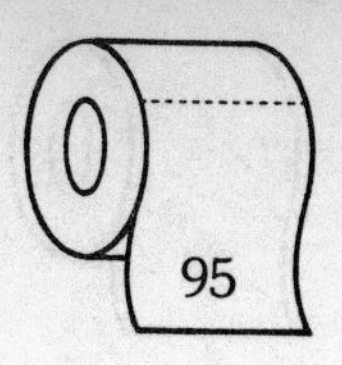

'Why municipal vandals should have thought it necessary to destroy so many (public urinals) I do not know; I suppose it is one expression of anti-homosexual prejudice. Yet no homo, cottage-cruising, ever prevented a hetero from merely urinating; while to do one's rounds of the cottages . . . provides homos, not all of whom are given to rougher sports, with healthy exercise.'

Tom Driberg

★ ★ ★

'You don't feel like some big rock star when you've got your head stuck down the toilet.'

Johnny Winter

★ ★ ★

'A certain necessary place has got the most lovely new pan you ever saw. It's quite a pleasure to look into.'

Wilkie Collins

★ ★ ★

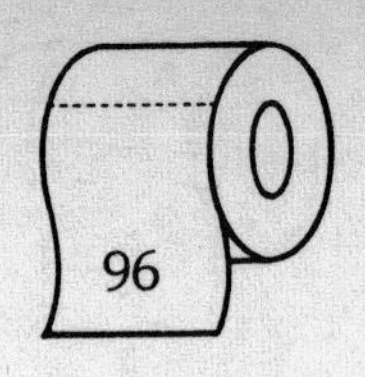

'We can piss anywhere.'
Mick Jagger

★ ★ ★

'When the loo paper gets thicker and the writing paper thinner, it's always a bad sign at home.'
Nancy Mitford

★ ★ ★

'I have no need of such things.'
Winston Churchill (to a plumber about to put a seat on his toilet).

★ ★ ★

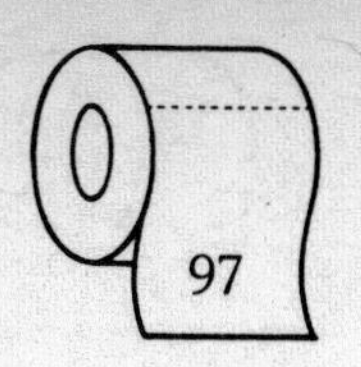

Liable to Deprave and Corrupt

For many years, the Lord Chamberlain kept all sorts of disgusting lavatorial references off the British stage. Among his most memorable instructions were:

1858 – From *Winter in Torquay*, first page, omit 'constipating'.

1957 – For 'ass upwards', substitute 'cock-eyed'.

1958 – For 'I'd like to pee', substitute 'I'd like to relieve myself'. And for 'What about that pee?' substitute 'What about that relieving yourself?'.

1963 – From *The Bed-Sitting Room*, omit 'crap' and substitute 'jazz'. In 'The Daz Song', omit 'You get all the dirt off the tail of your shirt', substitute 'You get all the dirt off the front of your shirt'. Omit the chamber pot under the bed.

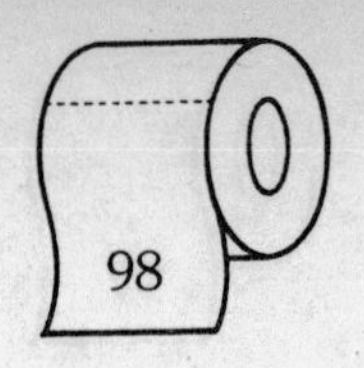

Members Only – Some news from London's most exclusive clubs

F. E. Smith, the first Lord Birkenhead, was in the habit of popping into the Athenaeum to make use of their members' lavatory. Having done this for years, he was somewhat surprised to be stopped, albeit politely, by one of the porters. The man discreetly informed him that he was in a private club of which he was not actually a member.

'Good God,' said Birkenhead, 'It's a club too, is it?'

★ ★ ★

A member of White's had invited an old friend up from the country and they were to lunch together at his club. As they walked through the streets of Mayfair, the countryman, remarking on the number of dogs' messes on the street, wondered how anyone could bear to live in London.

'What on earth makes you think they're dogs?' came the reply.

★ ★ ★

Two elderly members of a well-known club were relieving themselves in the recently redecorated loo of the club.

'What d'you think of this place now?' said the first.

'All right, I suppose. Needed a lick of paint.'

'Mmm,' came the thoughtful reply. 'Don't half make your cock look shabby, though.'

★ ★ ★

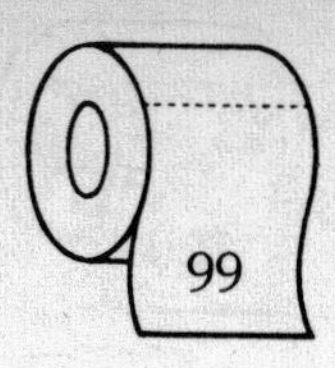

After a late sitting at the House of Commons – said by some to be the greatest club in the world – Winston Churchill entered the members' lavatory to find that he was alone with his great political rival Clement Attlee.

Noticing that Churchill had chosen to use the urinal that was furthest away from him, Attlee said, 'Feeling stand-offish today, are we, Winston?'

'That's right.' said Churchill. 'Every time you see something big, you want to nationalise it.'

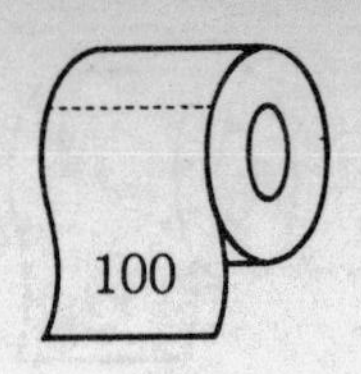

The Privy Thoughts of Pliny The Elder (AD 23-79)

Whoever makes water where a dog has previously watered will be susceptible to numbness in the loins.

Human urine cures the bite of a mad dog.

☆ ☆ ☆

A lizard drowned in urine has the effect of an antiphrodisiac upon the man whose urine it is. The magicians say the urine of a eunuch will have a similar effect.

☆ ☆ ☆

The urine of the newborn foal of an ass will thicken the hair.

☆ ☆ ☆

Goat's dung is an excellent cure for sore eyes.

☆ ☆ ☆

A plant upon which a dog has watered, torn up by the roots, and not touched with iron, is a very speedy cure for sprains.

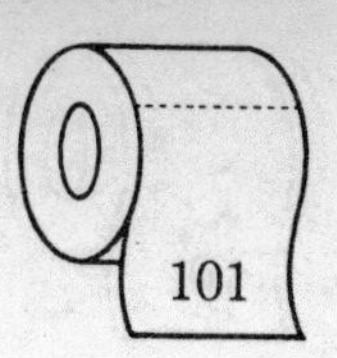

Apply bull's urine to the hair and it will rid the hair of lice.

Wolf's dung is a cure for cataracts.

For all diseases of the genitalia, wild ass's urine should be applied externally.

Pigeon's dung is an excellent gargle for those suffering from sore throats.

Those suffering from dysentery should drink the urine of a hyena.

Hen's dung will cure a bad case of flatulence.

The urine of a young boy will wash away freckles and improve the complexion of a woman.

☆ ☆ ☆

Crocodile dung cures epilepsy.

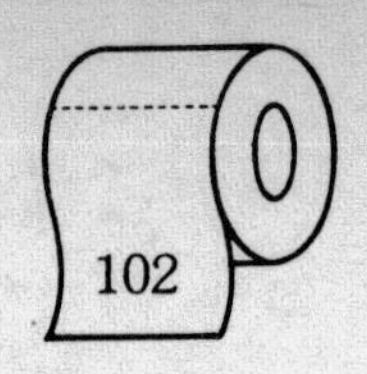

A particularly bad case of lice can be cured by application of human urine mixed with pigeon dung.

Those suffering from short-sightedness should consume chicken dung regularly.

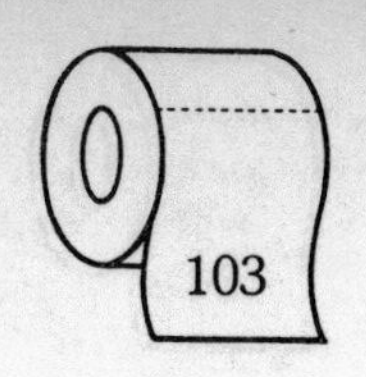

VI

NOW WASH YOUR HANDS (AND OTHER HINTS)

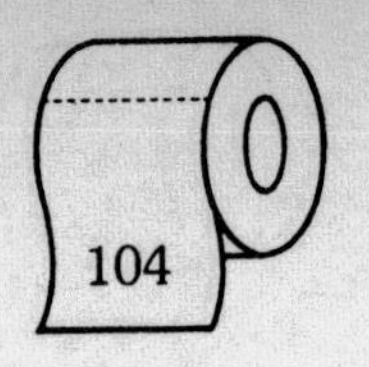

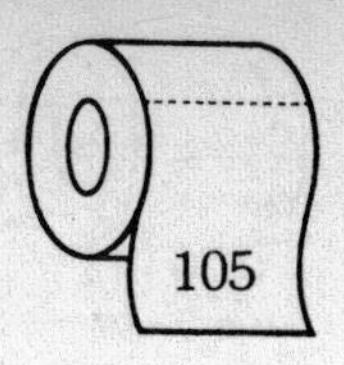

A gentle old dame they called Muir
Had a mind so delightfully pure
That she fainted away
At a friend's house one day
When she saw some canary manure.

★ ★ ★

There was a young lady whose joys,
Were achieved with incomparable poise.
She could achieve an orgasm,
With scarcely a spasm,
She could fart without making a noise.

★ ★ ★

There was a young man of Bengal
Who went to a masquerade ball
Arrayed like a tree
But he failed to foresee
His abuse by the dogs in the hall.

★ ★ ★

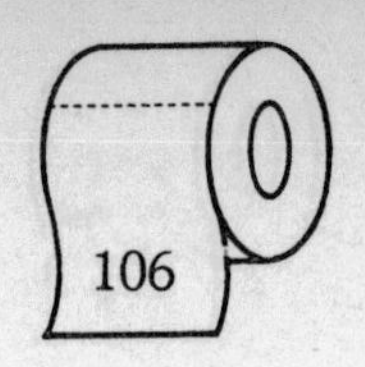

Public Lavatories – A Guide to Essential Etiquette

Nowhere is correct behaviour more important than in public lavatories. A careless remark, a suspect glance, or an ill-timed fart can reveal a person's lack of breeding and do untold damage to their position in society.

Yet, astonishingly, few if any of the style and etiquette books currently on the market address themselves to the question of toilet behaviour.

Here is a brief guide for the socially ambitious.

How To Be A Gentleman In The GENTLEMEN

Urinals

If, when you enter a public lavatory, there is one man using the urinals, do not on any account choose the one next to him. He will undoubtedly think you are an actor on the cruise. On the other hand, do not occupy the urinal furthest away from him since this suggests that you think he *is an actor on the cruise. Unless he is, this is regarded as unpardonably rude. When taking up your moderately distant position, do not look or smile at the other man. If a greeting seems appropriate, say 'Morning!' (or whatever) in a brisk, masculine voice and then ignore him.*

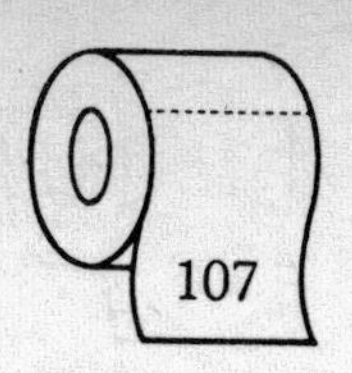

Adjacent urinals

When obliged to piss in close proximity to another man, do not look to see how he's getting on, however great the temptation. Stare straight ahead, as if reading a particularly interesting notice on the wall in front of you.

Undoing your flies

Do not make a performance of this. No one is impressed when you behave as if you're unpacking a six-pound trout, and to rummage around as if you've lost something merely provokes pity.

Pissing

If you are over ten years old, avoid playing the 'highest mark on the wall' game while relieving yourself. Similarly, it is unwise to direct your efforts at an angle directly parallel to the ground since this can splash yourself and others around you. The socially correct angle is 45° from the vertical, pointing downwards.

Holding

As complex a subject as the correct way of holding a knife at table, the actual hand position while pissing should be precisely that of a man watering his garden with a hose (but avoid waving it around). It is, of course, utterly unacceptable to hold it over rather than under, as if you have something to hide (even if you have). To use no hands at all is regarded as vulgar and may even suggest that you suffer from an intimate disease.

Farting

It is quite acceptable to fart, as long as it is not effeminate, half-hearted, whining or embarrassingly over-lengthy. A brisk, staccato effort is becomingly manly. Refrain from comment afterwards.

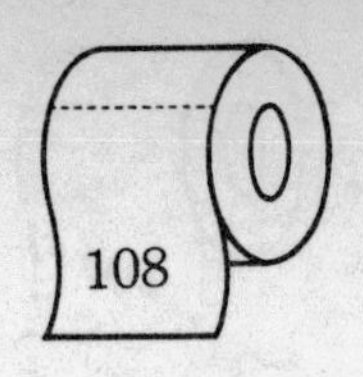

Wetting your trousers

To be avoided. If you have a 'drops around the crotch' problem, do not on any account try to cover it up by soaking the rest of your trousers at the sink. This fools no one.

In the crapper

A less complex operation since, except on rare occasions which are outside the scope of this book, it tends to take place in solitude, so the question of correct behaviour is less important. Avoid excessive grunts and never whistle or sing in a pathetic attempt to drown the noise. A harmless and agreeable sport is defacing the wall with graffiti, so long as what you write is suitably erudite and correctly spelt. Latin limericks are acceptable.

Interruptions in the crapper

Do not respond if someone is ill-bred enough to hammer on the door. Similarly, ignore pleas from the next cubicle for paper. This is an old trick – as you pass the paper under the partition, your hand will be grabbed and only released after your watch has been removed. Do not, in rougher establishments, succumb to the temptation of looking through the little holes into the next cubicle that some lout has drilled in the wall. There is always the possibility that you may find yourself eyeball-to-eyeball with a complete stranger.

Washing your hands

A brief rinse is all that's needed. A bigger splash suggests that you've been up to something in there – hence the film of the same name.

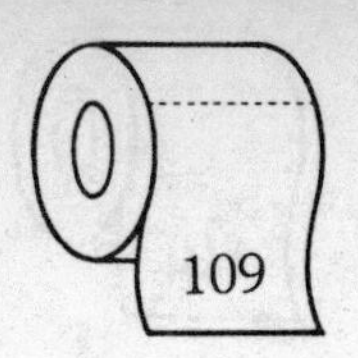

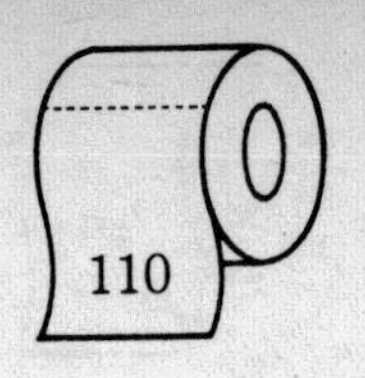

How To Be A Lady In The LADIES

Queuing

Always behave as if you were queuing for a theatre seat or a taxi. Chat amiably to all those around you, concealing your desperate need. Avoid hopping from foot to foot, crossing your legs, and gasping, 'For God's sake, hurry up *in there!'*

Urinals

Avoid using the urinals.

At the basin

Always wash your hands before and after going to the loo and generally fuss about in front of the mirror for as long as possible.

In the loo

Always aim for the side. The Niagara approach is generally regarded as being un-ladylike. If an audible fart escapes you, always go 'Whoops!' and giggle a bit. Never on any account have a crap in a public lavatory. There are some things that ladies should not do in public, and, this is one of them.

Passing the time

It is important to spend as much time in the loo as possible, since this helps reinforce the feminine mystique. Some ladies take a book, some write letters, others do their VAT returns. Many of the best of our contemporary novels have reached their first draft stage in the loo. If none of these appeals, resort to graffiti, the filthier the better. It is seen as being rather classy to leave the numbers of jilted lovers, particularly if they are cabinet ministers.

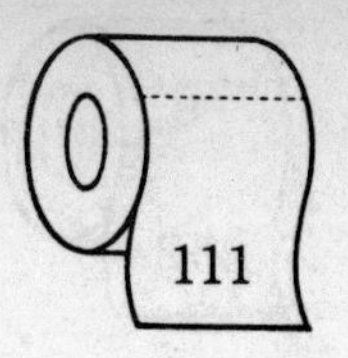

Afterwards

Again, spend some time in front of the mirror, but do not overdo the repair work to the face. Some ladies have been so effective in 'putting on their faces' that they have been completely unrecognisable and have had to re-introduce themselves to their party.

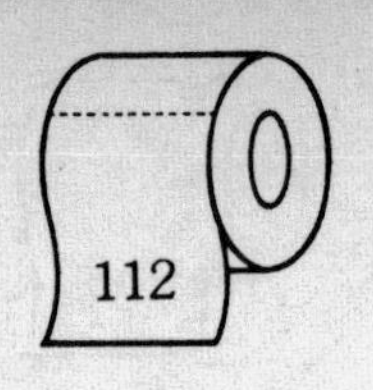

On The Board

Some public notices

'TOILETS OUT OF ORDER
Use platforms 3-4, 7-8.'

Notice on a railway station board

'FOR SANITATION PURPOSES PLEASE SANIT IN THE POTS PROVIDED AND NOT ON THE FLOOR.'

Sign in the Ladies of a Chinese Restaurant in India

'JUST IN – ARGENTINIAN COMBAT SUITS (TROUSERS SLIGHTLY SOILED).'

Notice in the window of an army surplus store in Diss, Norfolk during the Falklands War

'A BASEMENT FLAT COMPRISING 3 ROOMS, KITCHEN, BATHROOM, OUTSIDE W.C. (AT PRESENT OCCUPIED BY OWNER).'

Leaflet seen in an estate agent's window

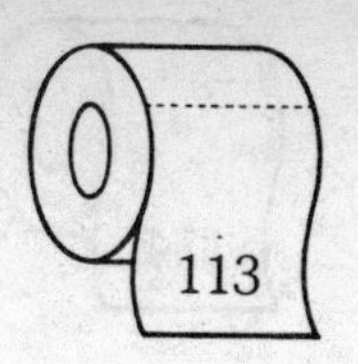

Toilet Behaviour – The Ayatollah Speaks Out

In his book *Political, Philosophical, Social and Religious Principles*, the Ayatollah Khomeini has laid down some very helpful principles for correct behaviour in the toilet. They include:

1. No one should urinate or defecate either facing towards or turning away from Mecca, in a public or holy place, facing the sun or moon, in a place exposed to the wind, under a fruit tree, or on a doorstep.

2. During the act, one should cover one's head, supporting one's weight on one foot.

3. One should also take care that one's genitals are hidden from the eyes of anyone who has reached puberty, including one's mother or sister, or anyone who is feeble-minded. One hand will suffice.

4. It is necessary to wipe yourself with one stone or one piece of cloth but it is important not to wipe oneself on a bone or any sacred object.

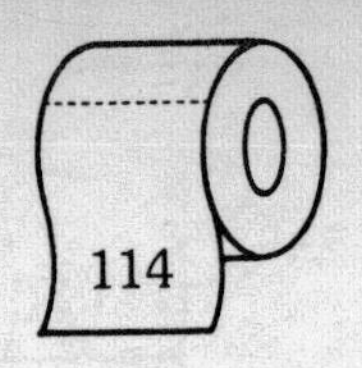

Vulgar, Obscene and Impure

Some lines in praise of euphemisms

Banish the use of the four letter words
Whose meanings are never obscure.
The Angles and Saxons, those bawdy old birds,
Were vulgar, obscene and impure.
But cherish the use of the weak-kneed phrase
That never quite says what you mean;
You'd better be known for your hypocrite ways
Than as vulgar, impure or obscene.
When nature is calling, plain speaking is out.
When ladies, God bless 'em, are milling about,
You may wee wee, make water or empty the glass;
You can powder your nose; even 'Johnnie' may pass;
Shake the dew off the lily, see the man 'bout the dog,
Or when everyone's soused, It's 'condensing the fog'.
But be pleased to remember if you would know bliss,
That only in Shakespeare do characters ——.
When your dinners are heart with onions and beans,
With garlic and claret and bacon and greens;
Your bowels get busy distilling a gas
That Nature insists be permitted to pass.
You are very polite and try to exhale
Without noise or odour (you frequently fail);
Excepting a zephyr, you usually start,
For even a deafer would call it a ——.
You may speak of a 'movement' or sit on a seat,
Have a passage, or stool – or simply excrete,
Or say to the others, 'I'm going out in back'
And groan in pure joy in that smelly old shack.

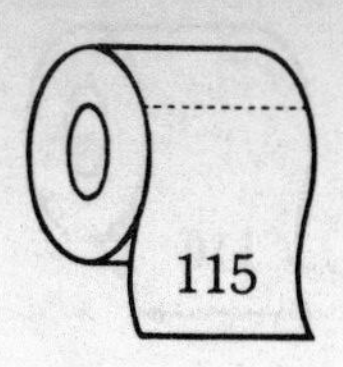

You can go 'lay a cable' or 'do number two'
Or sit on the toidey and make a 'do do',
But ladies and gentlemen who are socially fit
Under no provocation will go take a——.

So banish the words that Elizabeth used,
When she was a Queen on her throne.
The modern maid's virtue is easily bruised
By the four letter words when used all alone.
Let your morals be loose as an alderman's vest
As long as the language you use is obscure:
Today not the act but the word is the test
Of the vulgar, obscene and impure.

Poem by an anonymous author, quoted in W. S. Baring Gould's *The Lure of the Limerick*, from *The Limerick*: A Facet of Our Culture by A. Reynolds Morse.

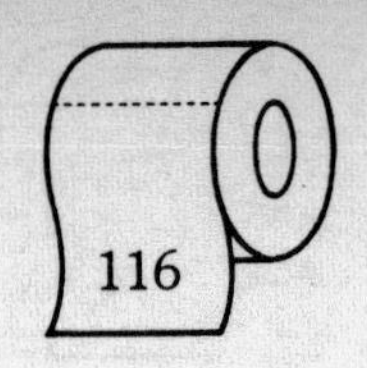

Books do furnish a room . . .

When equipping your small room, you want to be sure to find objects that fit in with your surroundings. Here are some suggestions of the sort of items that will reflect your state of mind.

Books

Peter Pan
Wind in the Willows
Great Expectations
Winnie the Pooh
Widdle of the Sands
Fair Stood the Wind for France
Gone with the Wind
Any work by T. S. Eliot – especially *The Wasteland* (plus a few titles published by Pan Books)

Pictures

Any work by:
Toulouse Lautrec
Whistler
Botticelli

Portraits

W. C. Fields
Flush Gordon
Trotsky
Ethelred the Unready
Richard the Turd
Stalin (on account of his purges)
Bronco (the horse)
Basil Brush

Historical etchings

The Relief of Ladysmith
The Battle of Waterloo

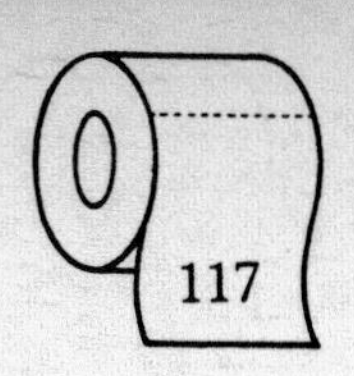

Framed maps

Windermere
Bogside
Pisa
Luton
The Windward Islands
Windy City
The Urals
Arsenal
Shit Creek

Music

Handel's Water Music
Any musical movement
Any work by Poulenc
Any work by Puccini

Pop

Any recording by U. B. Farty
Any recording by The Small Faeces
Penny Lane
I Can't Get No Satisfaction
(Who wants) Yesterday's Papers
I Can't Keep It In
I Hear You Knockin' (but you can't come in)
Where Do You Go To, My Lovely?

Newspaper

The Reporter

Political rosette

SDP only

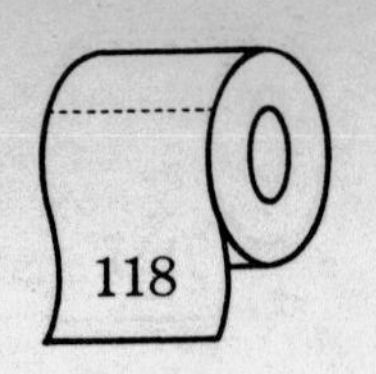
118

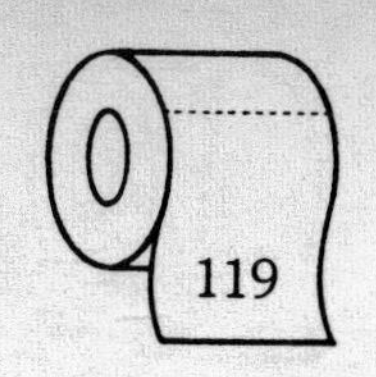

VII

SOME THINGS YOU'D RATHER NOT KNOW

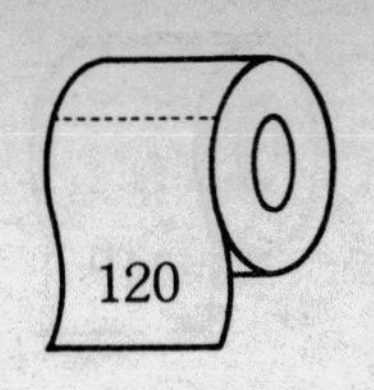
120

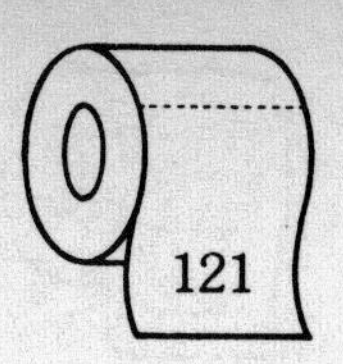

There was an old man who could piss
Through a ring – and what's more never miss.
People came by the score
And bellowed 'Encore!
Won't you do it again, sir? Bis! Bis!'

There was a young lady called Alice,
Who peed in a Catholic chalice.
The padre agreed
'Twas done out of need,
And not out of protestant malice.

There was a young man of Newcastle,
Who tied up a shit in a parcel,
And sent it to Spain,
With a note to explain,
That it came from his grandmother's arsell.

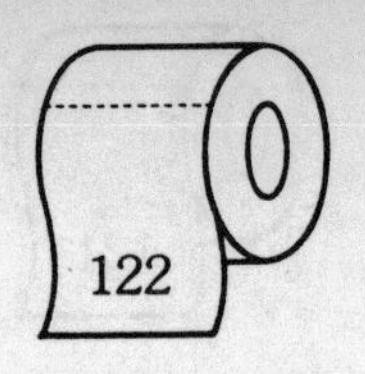

Messy Foreign Habits

Among the Tchukchi tribe of Siberia, it is customary for the guest of a family to be offered the daughter or wife of the host as a sign of hospitality. However, as a preliminary, the guest is offered a cup of the woman's urine with which to rinse out his mouth. If he refuses to drink it he is deemed an enemy of the family.

★ ★ ★

In an ancient rite, Hottentot boys are copiously pissed upon by the resident priest. They then have their left testicle cut off.

★ ★ ★

In ancient times, the Chinese had a complex form of punishment for lawbreakers. They were placed in a barrel or box filled with building lime. The box was then placed in the sun. The victim was encouraged to eat salty food and then given a lot of water to slake his thirst. As he relieved himself, the chemical reaction of the urine with the lime would burn him painfully to death.

★ ★ ★

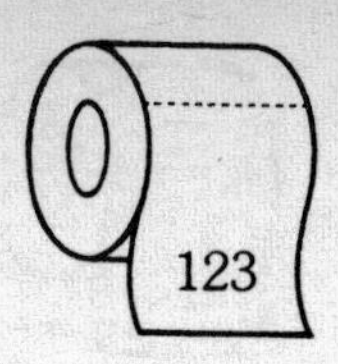

'On my first meeting with Mahatma Gandhi, being at that time a guest in his ashram near Ahmedabad, I had no sooner exchanged with him a few words of greeting than I was asked how were my bowels.'

Reginald Reynolds, *Cleanliness and Godliness*

★ ★ ★

According to the ancient laws of the Manu tribe in Asia, every conceivable part of the anatomy was liable to be amputated as a punishment. It was thought particularly important that the sentence should fit the crime. A man who broke wind in front of the king was likely to have his anus amputated.

★ ★ ★

When the nineteenth-century explorer and scholar Rockhill returned from Tibet to America, he brought back some pills much valued among the Tibetans. They were made from the excrement of the Lama and were widely used in Tibet and China as a cure for gun or sword wounds.

Soon after Rockhill had sent the pills for analysis, he received a letter from his friend, a Dr Mew, who wrote, 'I have at length found time to examine the Grand Lama's ordure, and write to say that I find nothing at all remarkable in it.'

★ ★ ★

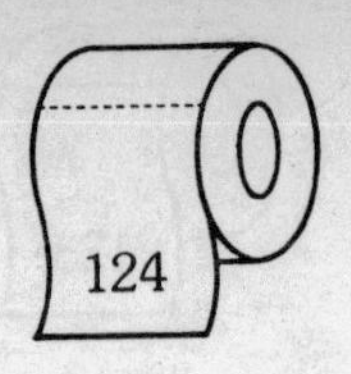

Among the country folk of nineteenth-century Maryland, a gruelling custom existed to test the love of a boy for a girl. When a father noticed his son becoming fond of a young girl, he would try to obtain some of her excrement. He would then make his son wear it under his armpit for a number of days. If the youth was still attracted to the girl after this test, he would be allowed to marry her.

★ ★ ★

Early explorers in Africa revealed that at Hottentot weddings it was customary for the priest to wander among the guests until he found the bride, urinate on her for a while, then set off in search of the groom, do the same to him, then return to the bride to repeat the performance until he was entirely spent. Priests would prepare for the ritual for days in advance.

According to Hindu law, a Brahmin must go through a lengthy washing process after any kind of bodily defilement. This should include washing his mouth out four times after urinating, eight times after defecating, twelve times after touching food and sixteen times after sexual intercourse.

A Berlin storekeeper was brought to court in the 1880s for having used the urine of young girls to make his cheeses richer and more piquant. Despite the scandal, there was an unprecedented demand for the cheese.

★ ★ ★

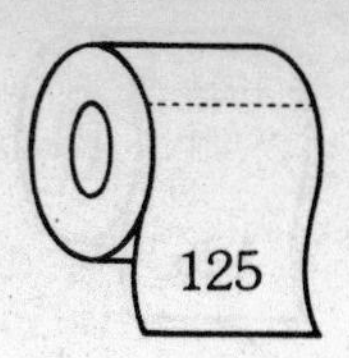

An American captain was given a love potion by a prostitute, after which he became obsessed with her. In order to break what they believed was an evil spell, his friends obtained some of her excrement and placed it in a new shoe. After the captain had walked around in the shoe until he could bear the smell no longer, he found to his astonishment that he was completely cured.

★ ★ ★

'Having list at any time to ease themselves, the filthy lousels had not the manners to withdraw themselves further from us than a Beane can be cast. Yea, like vile slouens, they would lay their tails in our presence, while they were yet talking to us.'

Friar William de Rubruquis, Louis IX's ambassador to the Grand Khan of Turkey, 1235

★ ★ ★

The great explorer Mungo Park received an unexpected and not entirely welcome honour one night in Africa. Sleeping in a village where a marriage ceremony was taking place, he was awoken by an old woman carrying a wooden bowl, the contents of which she threw in his face. It was a present from the bride – some of her urine. Park was told that this was a mark of distinguished favour.

★ ★ ★

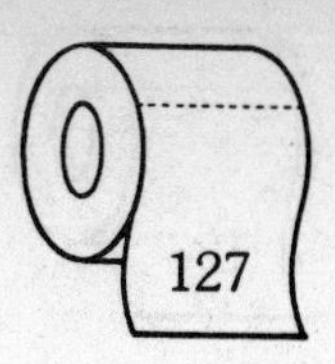

People living in East Siberia were said by travellers to keep a great pot of the family's urine in which they would all bathe. The sediment gathered around the top of the pot would be smeared on the body to kill vermin.

★ ★ ★

For Muslims, it is essential that urine does not fall on their clothes and defile them. They therefore squat to urinate and, having finished, they squeeze themselves carefully and sometimes dry the end against a wall. In his Voyage au Levant, *the French writer Tournefort writes, 'Sometimes one may see the stone worn in several places by this custom. To make themselves sport, the Christians smear the stones sometimes with India pepper . . . and some other hot plants, which frequently causes an inflammation in such as happen to use the stone. As the pain is very smart, the Turks commonly run for a cure to the very Christian surgeons who were the authors of all the mischief. They never fail to tell them that it is a very dangerous case, and that they should be obliged, perhaps, to make an amputation.'*

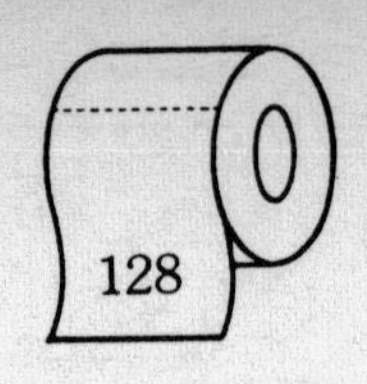

A Dirty Old Woman in London Did Dwell

A dirty old woman in London did dwell
That dirty old woman I knew her so well
She went to a doctor for she couldn't shite
And he gave her a pill which was like dynamite
It was brown brown dirty old brown.

That dirty old woman she came home to bed
She jumped on her knees and she stood on her head
She stretched for the po but the po couldn't grasp
So she upped with the window and popped out her arse
It was brown brown dirty old brown.

A London policeman was walking his beat
Was walking his beat at the end of the street
He looked up above at the stars in the sky
When a bloody great turd hit him slap in the eye
It was brown brown dirty old brown.

That London policeman he cursed and he swore
He called that old woman all sorts of a whore
On the London Bridge now you can see him by night
With a card round his neck 'I was blinded by shite'
It was brown brown dirty old brown.

A Cockney ballad quoted in Stranks' and Grant's Are You Sitting Comfortably

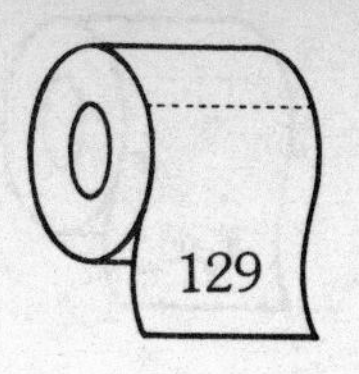

C'Est Incroyable!

The French and their toilettes

The appalling laxity in the toilet department of the entire French nation is legendary.

None of the following facts is exactly true but, if enough people start mentioning them to their friends, they'll be in the record books in no time . . .

The flush toilet was not introduced to France until 1949.

Napoleon never wiped himself, regarding it as effete and ladylike.

Four out of five French family homes has the old-fashioned hole-in-the-ground loo.

During the cold winter of 1971, nineteen people in Paris were admitted to hospital having been bitten by starving sewer rats who, driven to the surface by the cold, had leapt out of the traditional squatting loos and savaged the lower parts of those using them.

It is not regarded as *comme il faut* in the best French households to wash after having a satisfactory time in the toilette.

In Alençon, it is legal to do *numero deux* in the street pissoir – even if you're a woman.

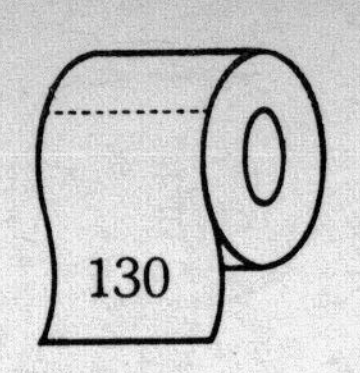

The similarity between *la mer* (the sea) and *la merde* (shit) is not accidental. It is traditional for the French to do it in the sea while *en vacance*.

Blaise Pascal was one of the first overseas visitors to enjoy Sir John Harington's flush toilet. The experience was said to have prompted one of Pascal's most famous *Pensées*, '*Apres moi, le deluge*'.

Le Petomane, the famous farter, was not really French. Born in Brighton as Eric Crowhurst, he discovered that no one in England was particularly interested in a man who could hit top C with a fart. He emigrated to France, and made his fortune.

Over 87% of all medicines taken by the French is in the form of suppositories.

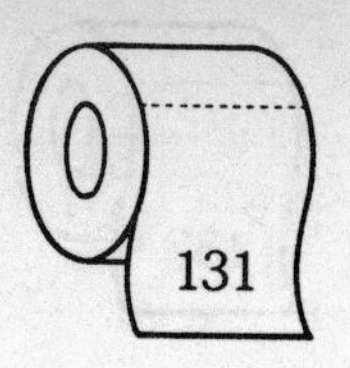

Hold the Front Page

'You are never more than four minutes from a public convenience in the City of London,' said the spokesman proudly, 'providing you run.'

Daily Telegraph

'The librarian at the Building Research Station was somewhat puzzled by the demand for old copies of the *Daily Telegraph*. This too from the plumbing section who seemed to be showing an unexpected interest in right-wing politics. It emerged eventually that the standard test of a good flush is the ability to swallow half a *Daily Telegraph*.'

New Scientist

'Despite furious banging on the door, the occupant of the train's only toilet refused to come out. Saragwa Salani stayed put, ignoring all threats and entreaties, for the whole of a 100-mile journey from the Tanzanian border to the Kenyan town of Voi. There, he was finally dislodged from the loo, arrested, charged with 'refusing to move out of a place reserved for public use' and fined the equivalent of £5. He said, 'I had never travelled in a train before and I was nervous.'

Weekend

'The judge will have the power to order off any dog that commits a fragrant error.'

Gowerton and District Annual Show Schedule

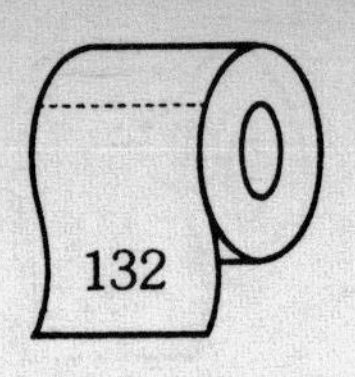

'At a time when beer drinking has risen by a third over ten years, the number of loos has declined by a similar fraction. The strain is beginning to tell as owners of front gardens in Earls Court frequently complain.'

Daily Telegraph

'A *phocaena phocaena* was found today propped up in one of the cubicles of the men's lavatory in Glasgow Central Station. The staff thought it was a dolphin but the four foot, sixty-four pound carcase was identified as a porpoise by the museum's department at Kelvingrove Park. How it had come into the lavatory nobody knew. One gentleman said. 'We had rain and there was flooding but this is ridiculous.'

The Times

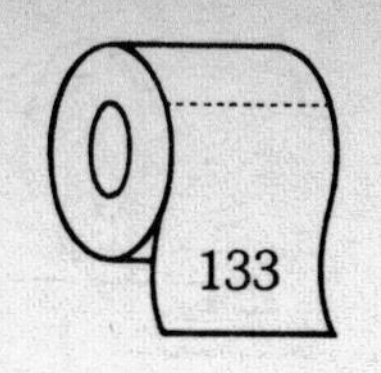

VIII

IT'S BREAD AND BUTTER TO ME

On a picnic a Scot named McFee
Was stung in the balls by a bee.
He made oodles of money
By oozing pure honey
Each time he attempted to pee.

☆ ☆ ☆

Said a printer pretending to wit,
'There are certain bad words we omit.
It would sully our art
*To print the word f****
*And we never, but never, say s***.'*

☆ ☆ ☆

There was a young man of Australia
Who painted his bum like a dahlia.
The drawing was fine,
The colour divine,
The scent – ah! that was a failure.

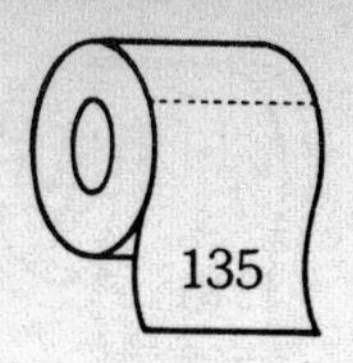

Business as Usual

The toilet has not always been a place where you could escape from the cares of the world for a few minutes' quiet contemplation. For some, it has been a great place for public as well as private business . . .

☆ ☆ ☆

During the last illness of Louis XIII, Anne of Austria would always accompany the King to the Garderobe where she would perform 'humiliating functions' for him.

☆ ☆ ☆

It was customary for the French court to be in attendance while Louis XIV was at stool. Madame de Maintenon would normally wait in there, while the King attended to affairs of state.

☆ ☆ ☆

While travelling around Europe in 1799, Sir N. William Wraxall stayed with Sir William and Lady Hamilton in Naples. There he observed the court of King Ferdinand IV. He was shocked by what he found.

'Those acts and functions which are never mentioned in England and which are there studiously concealed, even by the vulgar,' he wrote, 'here are openly performed. When the King has made a hearty meal, and feels an inclination to retire, he commonly communicates that intention to the Nobleman around him in

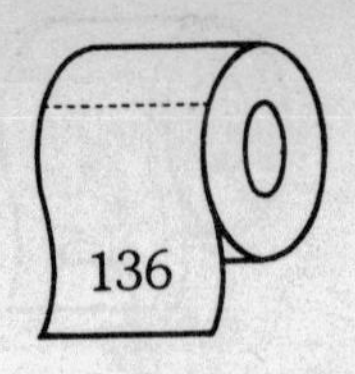

waiting and selects the favoured individuals, whom, as a mark of predilection, he chooses shall attend him . . . The persons thus preferred, then accompany his Majesty, stand respectfully round him, and amuse him by their conversation, during the performance.'

☆ ☆ ☆

To this day, according to Auberon Waugh, Lord Gnome the legendary editor of *Private Eye*, will only receive his minions while he is busy on the editorial throne.

'Lord Gnome receives me in the usual place. Marmaduke Gnome, I should explain, is one of the most remarkable men who have ever lived. Intensely shy, he can talk to his subordinates only when he is seated on the lavatory. He seldom looks at them and never refers to them by name – with the one exception of myself, who for some reason he always calls "Peregrine".

'Kneel down, Peregrine,' he says kindly as I am ushered in. A long silence follows. Plop. He says that he has received a letter from lawyers representing Dame Harold Evans, whose untimely death at the age of 91 I recorded last week.

"Dame Evans is dead," says Marmaduke. Plop. "You shall never again refer to him, by name. Nor to his house, to his wife, his manservant, his maidservant, his ox, his ass, nor to anything that is his." Plop. Plop. Plop.'

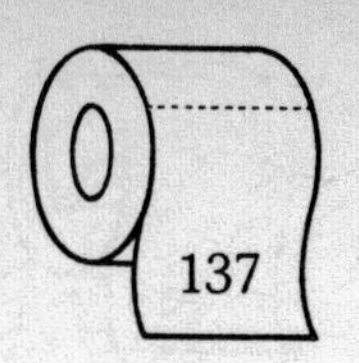

The Hard Sell

Some revealing moments from the personal and small ads columns

'Lonely lady, 43, with little dog, seeks post.'

Express and Echo

'You could have a portrait of yourself or your child taken at the convenience of your own home.'

The Times

'Hall with kitchen and toilet (can seat 100 persons) or could be converted for dwelling house.'

Edinburgh Evening Post

'PLUMBING. New installations. Repairs to old approved by Age Concern.'

Leicester Mercury

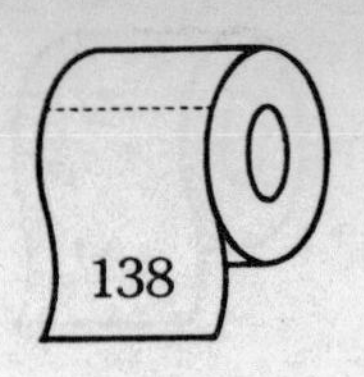

Household Hints from the Smallest Room

Over the centuries, the many unusual properties to be found in the 'nameless ingredients' that we ourselves can produce have been expounded by conservation-minded individuals. Surely, in this ecologically conscious age, there's a place for some of these?

During the nineteenth century, much of the best quality tobacco was cured in privies above the fumes of human urine.

★ ★ ★

The Chinese are known to have used human dung to hatch eggs.

★ ★ ★

In mediaeval France, a variety of aphrodisiacs could be concocted by women anxious to stir the lust of a man. They could give him a cake containing some solid contributions from their own privy (which, alternatively, they could place in his shoe for the same effect). Otherwise, the same ingredients put in a man's porridge were said to work a treat.

★ ★ ★

To this day, natives of a Pacific South Sea island squat over the end of a pier into the sea and catch the fish that swarm in for the food.

★ ★ ★

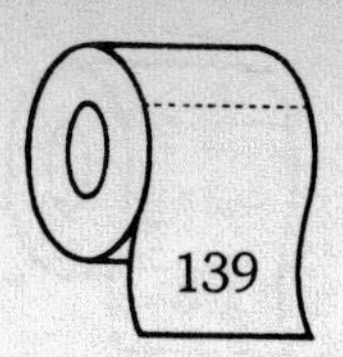

Eskimos use urine as a shampoo.

★ ★ ★

Urine was regarded as such a precious substance in Roman times that a tax was levied on its sale.

★ ★ ★

In *Mary Lou's Household Hints*, sometimes known as 'the Mrs Beeton of the bathroom', we learn of several uses for urine. It can be mixed with bracken ashes and used to polish pewter; or, fermented, it will remove grease from household linen; applied to the face, fresh every morning, it acts as a fat solvent and clears the pores; finally, if you have a fair hair, a few drops sprinkled on and allowed to dry in the sun will result in fashionable and attractive streaks.

★ ★ ★

Less recently, it was known for women in England to drink their husbands' urine when they went into labour. It was said to make giving birth easier and less painful.

★ ★ ★

In the Middle Ages, a urine bath was known as being good for the feet and an excellent preservative against the plague.

★ ★ ★

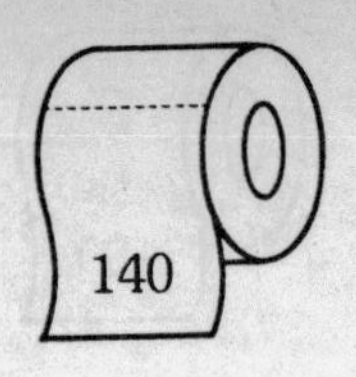

Characters called Piss Dick or Piss Joe used to be familiar characters in Welsh villages. They would come round to buy urine which they would collect in two large pails slung over a horse. The payment was 1d a bucket, except for redheads, who were thought to have different kidneys. They would get 1½d.

★ ★ ★

Until recently, a cure for deafness was for someone to pee into the afflicted person's ear.

★ ★ ★

WAAF's and nurses used to stand their new boots in chamber-pots to soften the leather.

★ ★ ★

Schoolmasters, who believe in the traditional form of discipline, are wont to use urine to harden the cane for the next beating.

★ ★ ★

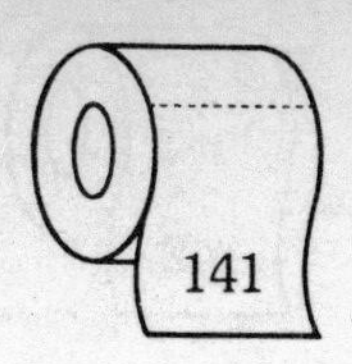

A Medical Hint from India

'I possess irresistible medical statistics to prove that rhinoceros urine is a cure for the common cold. It was the British who crushed our natural ayurvedic health system. People say that I am a Hindu fanatic. This is nonsense. I drink brandy and smoke several cigarettes a year.

Every Indian family should keep a bowl of urine, preferably very old urine, on the kitchen shelf. Urine knows no religion.

Why only the other day, when I was flying from Calcutta to Delhi, I engaged a fellow passenger in conversation. He turned out to be C. Szent Gyorgagee, India's Poet Laureate, who was suffering from English 'Flu. He took a deep swig from my bottle of Rhinorine and by the time we landed he was cured.

What further evidence is needed to show that my treatment is the only treatment?'

Mr Jagdish Morarji, the prominent urinobibe, reported in the Bombay Standard

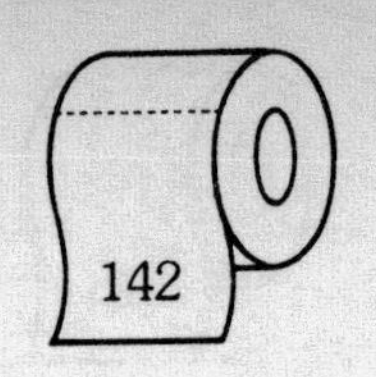

Up The Wall – some graffiti that has made its mark

IF YOU CAN AIM THIS HIGH YOU SHOULD BE IN THE FIRE BRIGADE

PRUNE JUICE SETS YOU FREE

NO BOY CAN SWIM EXCEPT IN WATER PASSED BY THE HEADMASTER

I THOUGHT WANKING WAS A TOWN IN CHINA UNTIL I WENT BLIND

THE ONLY DIFFERENCE BETWEEN PHILOSOPHY AND GRAFFITI IS THE WORD FUCK

AVE MARIA: DON'T MIND IF I DO

VD IS NOTHING TO CLAP ABOUT

I MUST BE A MUSHROOM. I'M KEPT IN THE DARK AND FED BULLSHIT

T. S. ELIOT IS AN ANAGRAM OF TOILETS

LIFE IS A HEREDITARY DISEASE

MORE DEVIATION, LESS POPULATION

JOIN NOW – THE HERNIA SOCIETY NEEDS SUPPORT

SNOW WHITE THOUGHT 7-UP WAS A DRINK UNTIL SHE DISCOVERED SMIRNOFF

HELP! I'M TRAPPED INSIDE A HUMAN BODY

ANTISOCIAL DISEASES ARE A SORE POINT

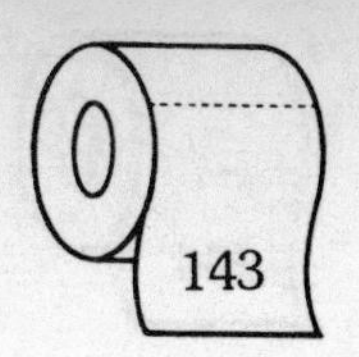

MEN ARE LIKE A PUBLIC TOILET – THEY'RE EITHER
VACANT, ENGAGED OR FULL OF SHIT

IT'S NO GOOD LOOKING UP HERE FOR A JOKE. YOU'VE
GOT ONE IN YOUR HAND

I REALLY DON'T LIKE THIS PLACE AT ALL
THE SEAT'S TOO HIGH AND THE HOLE'S TOO SMALL

(TO WHICH I MUST ADD THE OBVIOUS RETORT
YOUR ARSE IS TOO BIG AND YOUR LEGS ARE TOO SHORT)

IS REGENTS PARK TOILET A ZULU?

(under the light) A LIGHT TO LIGHTEN THE GENITALS

IT'S NO GOOD STANDING ON THE SEAT
THE CRABS IN HERE CAN JUMP TEN FEET

DIARRHOEA WAITS FOR NO MAN

DON'T THROW YOUR FAG ENDS IN THE LOO
YOU KNOW IT ISN'T RIGHT
IT MAKES THEM VERY SOGGY
AND IMPOSSIBLE TO LIGHT

ONE WOULD THINK TO READ THIS WIT
THAT SHAKESPEARE HIMSELF CAME HERE TO SHIT

Tongue-twister

The cat crept into the crypt, crapped, then crept quietly out.

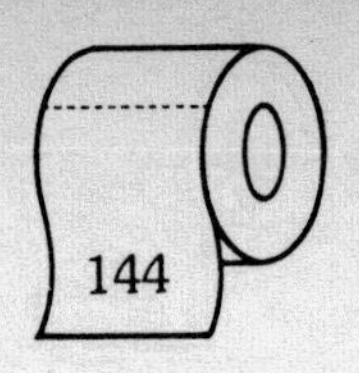

The catalogue

For American farming communities, the enormously thick mail order catalogue sent out to all farmers by the firm of Sears Roebuck was an annual supplement to the stocks of lavatory paper. The massive volume would be hung on a nail in the privy. Charles Sale, in the all time lavatorial bestseller *The Specialist*, ponders the longevity of this publication. The book's narrator is Lem Putt, whose speciality and love was building privies. 'I'll tell you about a technical point that was put to me the other day. The question was this: "*What is the life, or how long will the average mail order catalogue last, in just the plain, ordinary eight family three holer?*" It stumped me for a spell; but this bein' a reasonable question I checked up, and found that by placin' the catalogue in there say in January – when you get your new one – you should be into the harness section by June; but, of course, that ain't through apple time, and not countin' on too many city visitors, either.

An' another thing – they've been puttin' so many of those stiff-coloured sheets in the catalogue here lately that it makes it hard to figger. Somethin' really ought to be done about this, and I've thought about takin' it up with Mr Sears Roebuck hisself.'

Charles Sale
The Specialist, *1930.*

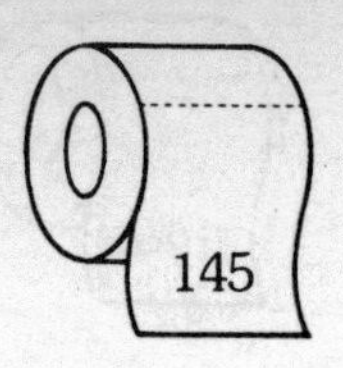

Toilet Tissue

A Perforated History

Today, when most historical activity has been fully documented by academics, the history of toilet tissue has been sadly neglected. Until now.

Early archaeological evidence tells us that the primitives were in the habit of using sticks and stones as toilet tissue, but by the time the Roman Empire was at its height, perfumed wool or small sponges on sticks had been substituted for the crude early devices. There is no evidence from the Dark Ages, when so many civilised habits were allowed to lapse, and one suspects the use of toilet tissue would be among those forgotten in these barbaric times. But with the dawn of the Middle Ages, we again find evidence of man's attempt to find himself a decent toilet tissue. Monks at an abbey in St Albans used torn-up squares of old habits in the privy and the mediaeval laity, depending upon their station, used carved sticks or hay.

Even when paper became a common commodity in the fifteenth century, there was not sufficient available for use in the privy. Books became a common source of toilet tissue, a habit that scandalised Sir Robert Herrick (1591-1677) who, in an attempt to crush this habit, inscribed the following lines in the front of his books:

Who with thy leaves shall wipe (at need)
The place where swelling piles do breed;
May every ill that bites or smarts
Perplex him in his hinder parts.

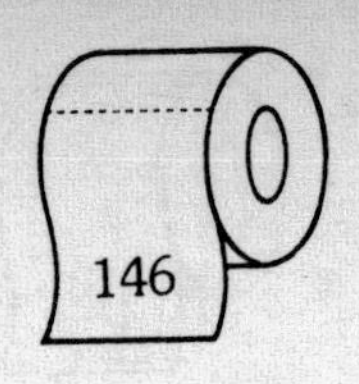

And of course correspondence, too, became a common source of tissue for the needy. 'I am seated in the necessary house,' an eighteenth-century correspondent was said to have written. 'I have your letter before me. Soon it will be behind me.'

Among the most bizarre items to be used for cleansing is the reported use of mussel shells in the eighteenth century. The author of a book on degeneracy writes: 'I have known an old woman in Holland set herself on the next hole to a gentleman and civilly offer him her mussel shell by way of a scraper after she had done it herself.'

And in America the corn cob was in frequent, if uncomfortable use.

The earliest published mention of specifically produced toilet paper is in a book published in France in 1718. The author tells how the Arabs travelling in China in the ninth century were confronted with such paper. Not being familiar with it, they used it for playing cards.

But it was not until 1871 that Mr Seth Wheeler, one of the forgotten benefactors of mankind, put the race out of centuries of misery by producing the first toilet roll. He took out a patent for a machine which could perforate sheets measuring 5 inches x 6 inches (12.5cm x 15 cm). But it was still not until the 1880s that such paper became available in Britain through the British Perforated Paper Company. The new product seemed to face almost insuperable difficulties, for in the strait-laced atmosphere of Victorian England it was impossible to mention it by name. Housewives were in the habit of asking for the new-fangled product under the euphemism of 'curl papers' or simply requesting of the understanding shopkeeper, 'I'll have two please.' Despite this handicap, the new product proved so popular that within twenty-five years it was used everywhere.

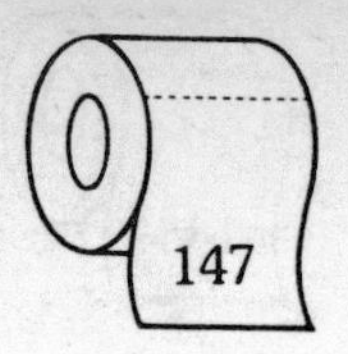

But even today toilet tissue is surrounded by controversy. Soft, two-ply tissue is blamed for the fact that tummy upsets are on the increase. Andrew Semple, formerly Professor of Environmental Health at Liverpool University, has condemned modern toilet tissue as being quite useless in preventing hands from being contaminated by germs. British scientists thought they'd found a solution to this problem when they came up with a new type of tissue. It felt and looked like the usual soft product but it contained a skin of 'polyvinyl alcohol' which formed a germ proof barrier. This tissue was briefly available in the shops, but the cost was so high – 69p for fewer than fifty sheets – that people just weren't prepared to splash out for it. But experts still believe toilet tissue will change as much over the next hundred years as it has over the last. Perhaps the technology of space travel will play some part . . . All we can say for sure is that at the moment no one knows what's in the pipeline.

Paperwork

Until early this century, newspaper squares were used in the lavatories at Windsor Castle. The equerry to the Prince of Wales, later to be Edward VII, wrote: 'We are all fairly comfortable in this most conveniently built house . . . We all admire various little economical thrifty dodges here. In the W.C.'s – the newspaper squares!'

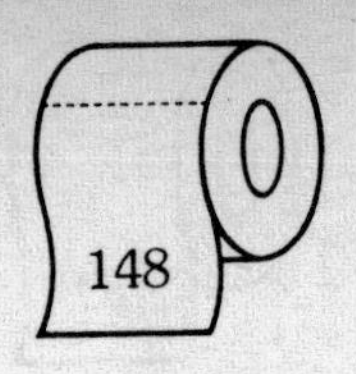

After unfavourable reviews of his work, American producer Alexander Cohen gave away loo paper to his friends printed with the name and photograph of an offending critic John Simon.

During Victorian times, untreated sewage still ran into the Cam. When visiting Cambridge, Queen Victoria was shown round Trinity by the master, Dr Whewell. Looking over a bridge, she asked, 'What are all those pieces of paper floating down the river?' Whewell replied with great quickness of mind, 'Those ma'am are notices that bathing is forbidden.'

Airlines are alleged to have introduced the compulsory use of toilet squares after the following unfortunate incident. A nervous passenger was occupying the small room of a DC3 when rough weather was encountered. The 'return to seat' sign came on, and the seated man, acting with more haste than discretion, dressed hurriedly and got the end of the roll tucked in his trousers. He fled down the gangway leaving a trail of paper behind him. The hilarity of the other passengers made it quite impossible for the steward and stewardesses to ensure that safety procedures were followed.

A West German toilet paper manufacturer has printed an entire English language course on his paper. There are twenty-six lessons on each roll, and to prevent family squabbles, the course is repeated eight times on each roll.

In 1971 the Mayor of Santhia in Italy ordered a large quantity of toilet paper for the municipal workmen. At the next council

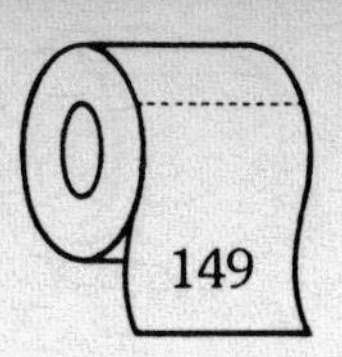

meeting he was accused of over-paying for the paper and in the course of a one hundred minute speech he defended his decision, lyrically praising the qualities of the paper. To prove his point, he distributed free rolls of the paper to each councillor so that they could try the paper for themselves. But it was all to no avail. The storm had leaked to the press, and the mayor was forced to resign, which meant under Italian law that the whole council had to be re-elected because of the bog paper controversy.

★ ★ ★

At Your Convenience

Some of us are content to use lavatories for the purpose for which they were installed. Others like to take things a little further . . .

Recording Studio

The pop group Slade recorded one of their LPs in a gents' lavatory when they found they couldn't create the sound they wanted in a heavily carpeted studio. The group reported that the public toilet gave their music new tone, but the venue wasn't without its problems. As Dave Hill, lead guitarist, said afterwards: 'We had to record it all one weekend, when other people didn't want to spend a penny.'

And the idea caught on. Matabolist, a South London heavy rock band, leased the men's underground lavatory at Lavender Hill, Battersea, for recording.

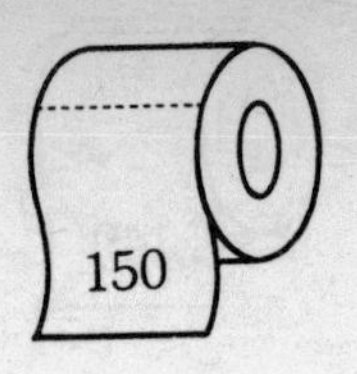

Flower pots

Old age pensioner Wally Cogan kept his geraniums in urinals outside his house until his landlords – the Church in Wales – ordered him to take them down because they weren't in keeping with the village's prize-winning image. The instructions were to remove them 'at his earliest convenience'.

Polling Station

Broadstairs County Council plans to convert a disused public toilet into a polling station.

Art gallery

Fred Ward uses a public convenience in Chatham, Kent, as an art gallery. He shows interested visitors round the convenience where one wall is covered by a twenty-foot mural, and the others are used to hang portraits. Fred sees nothing unusual in this arrangement: 'I regard it as a public building where people can come for a little education and enjoyment.'

Calendar

East Hertfordshire County Council used a picture of one of their public lavatories on their annual calendar. The council chief commented: 'The building is different. It's not an unpleasant picture.'

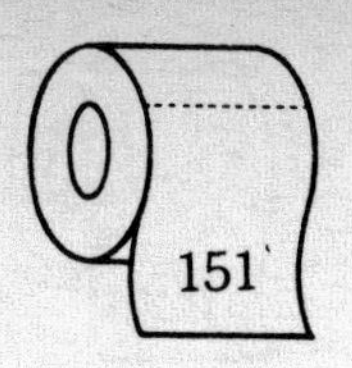

Santa's Grotto

Jean Hemmett turns her five-cubicle powder room near the Market Hall, Stockport, into a Christmas haven. Come the festive season, the walls are bedecked with twenty yards of streamers and reindeer nuzzle among the lavatories. Every child who visits is given a free sweet and a sight of Santa and a warm cuppa greets the elderly. Says Jean, 'It's a time of good cheer, and tinsel and toilets go well together.'

A home from home

Paul Goddard, freelance clown and gag writer, is trying to rent a derelict public lavatory from Hammersmith and Fulham Council. The handsome mock Tudor structure was built in the 1930s and fell into disuse when a modern new superloo was built just outside. Goddard is short on money but says he's good with his hands and reckons he will make short work of the conversion. The council is said to be sitting on the request.

Restaurant

Plans have been lodged with the town council in Reading, Berkshire, to turn Victorian public toilets into a tea room.

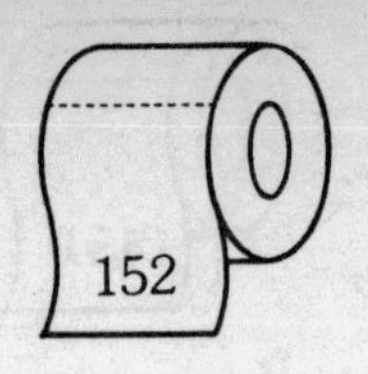

Jokes, Practical and Impractical

In his book *Gullible's Travels*, Billy Connolly reveals a handy trick with stink bombs and train lavatories:

'On the bottom of lavatory seats are four wee black rubber knobs. What we would do is put the stink bomb on one of those, tape it with black tape so it couldn't be seen, then hang around outside. The first person to go in would put the seat down and there would be this most disgusting smell. Then we would bang on the door and go: "Hurry up, in there . . . JESUS! What a smell!" And the poor bloke would be inside, unable to open the window yet not wanting to come out – he'd be stuck in there all the way to London . . .'

☆ ☆ ☆

During the 1940s, a Hollywood host had a gimmicky outside loo which he'd invite new guests to his house to use. But when they pulled the plug, the walls would collapse, revealing the luckless individual to the gaze of other guests who had gathered around for the spectacle.

Rather more easy to organise is one of the simplest and most disruptive jokes in the history of the small room. You simply cover the pan of the lavatory with cling film. Apparently the effect can be shattering.

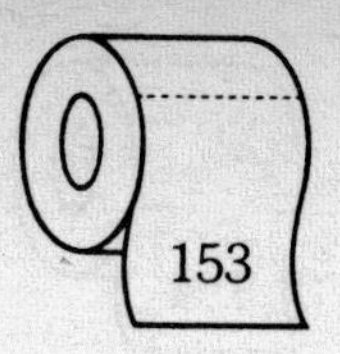

American pornographer Russ Meyer's favourite practical joke involves stealth, timing and a very long shovel. The idea is to follow someone who is trying to find a quiet place outside to do his business, having been caught short. As he squats down, you use the shovel with a long handle to catch his impressive efforts and quickly whisk them away. Having completed his task, the man will look round, as is natural, only to find that there is absolutely nothing to be seen. Soon he will be desperately searching everywhere for the lost turd – in his shoes, his turn-ups, the back of his trousers – all to no avail.

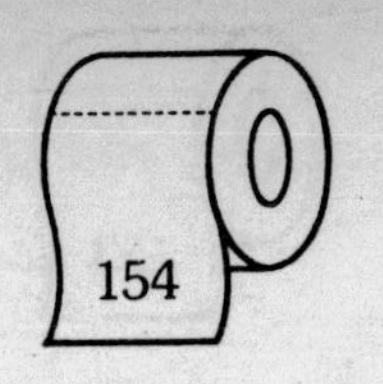
154

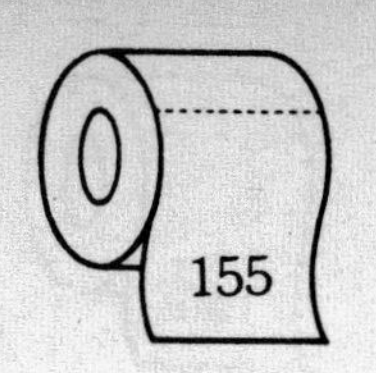

IX

FROM SLOW TO STOP

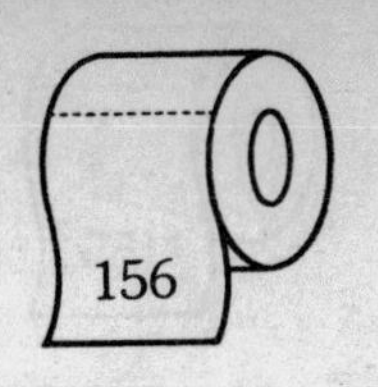

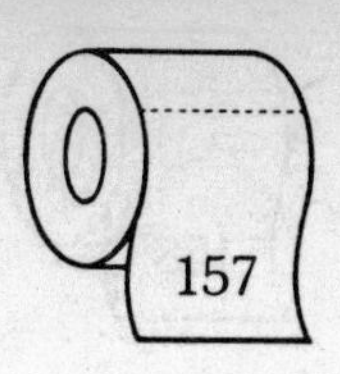

There was once a fellow named Howells
Had a terrible time with his bowels.
His wife, so they say,
Cleaned them out every day
With special elongated trowels.

There was a young lady at sea,
Who complained that it hurt her to pee.
Said the brawny old mate:
'That accounts for the fate
Of the cook, and the captain, and me.'

There was an old man of the Isles,
Who suffered severely from piles.
He couldn't sit down
Without a deep frown
So he had to row standing for miles.

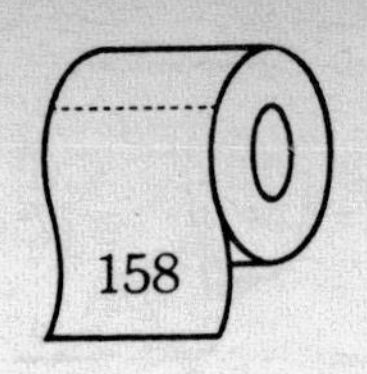

Slow Movements

Some important information about constipation

'Habitual constipation is a disease of civilization, and affects all classes and all ages. Although not fatal nor even a dangerous disease, it may drag on for many years, often for an entire lifetime, embittering existence, destroying happiness, and, in exceptional cases, making life intolerable.'

Professor Ismar Boas, *Habitual Constipation*, 1923.

Among the pioneers of the fight against constipation which, during the late nineteenth century, was seen as almost as great a threat to moral, physical and mental hygiene as masturbation, was a Viennese doctor who advocated a treatment involving repeated abdominal massage. His name was Dr A. Bum.

'The ill effects of excessive tea drinking are due to their influence upon the digestive and nervous systems. If taken in excessive quantities with meals, it retards digestion and usually produces constipation . . . Many a fretted, nervous housewife may thank her tea for her uncompanionableness.'

M. Wood Allen, *What a Young Woman Ought To Know*, 1901.

☆ ☆ ☆

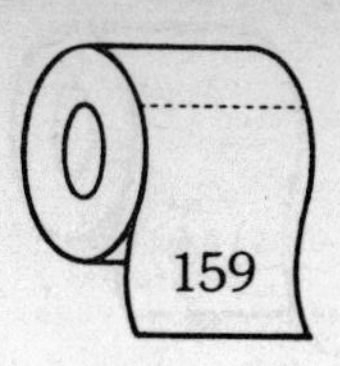

The current record for constipation is 102 days.

☆ ☆ ☆

The **Scientific American** ***has reported a major French breakthrough in the treatment of constipation. A woman patient of Dr Cabe of Lyons had been*** **sans mouvement** ***for over forty days. The doctor applied the negative pole of a Gaiffe battery to the lady's rectum and the positive to her navel and turned the current on.***

'Two minutes later,' reported the doctor, 'we had a satisfactory movement.'

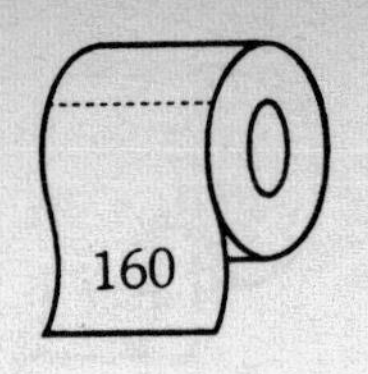

True or False?

The first three patents on water closets were taken out by a watchmaker, a cabinet maker and a cook.

(True. Alexander Cunningham, a Bond Street watchmaker, took out the first patent on the water closet in 1775. Joseph Bramah, a cabinet maker, took out a patent for the 'valve closed' in 1778 and John Gaillait, a cook, took out a patent for a 'stink trap' in 1782.)

Einfahrt ***and*** **ausfahrt** ***are German for ladies and gents.***

(False. Although many British motorway drivers in the Bundesrepublic have made this simple mistake to their cost.)

The word lavatory is a euphemism.

(True. It comes from the French word for washbasin.)

In Japan some factories are installing digital drying machines in place of loo rolls.

(False. Although it is a subject discussed annually and with much heat at the Conference of the British Institute of Personnel Directors.)

The word 'loo' derives from the Scots cry **gardez loo,** ***or look out for the slops which were thrown out of an upper window.***

(True. The cry *gardez loo* was in fact a corruption of the French *gardez l'eau*, another euphemism.)

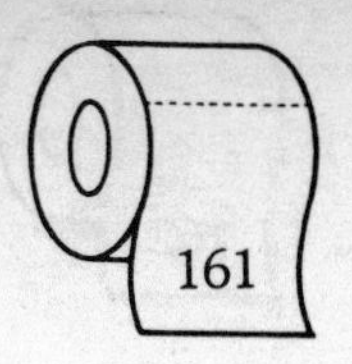

Undignified Exits

A man in Bavaria was found dead on his toilet with a potato masher attached to his member. The inquest was told that the man had adapted the electric masher for purposes of self-relief but, having attached it to the light socket in the lavatory, he made the mistake of touching the chain, thereby providing a perfect earth and electrocuting himself.

★ ★ ★

'At the Lancaster Assizes, Walter Moore was found guilty of murdering his wife at Black Lane-ends, Keighley, by cutting her throat with a razor. In this instance, the convict anticipated his doom by a few hours, committing suicide in a water-closet, by thrusting his head into the pan and letting on the water.'

The Times, August 8th, 1962

★ ★ ★

In 1589, Henry III of France gave an audience while sitting on the *chaise percée* to a monk called Clement, who took the opportunity to assassinate him.

★ ★ ★

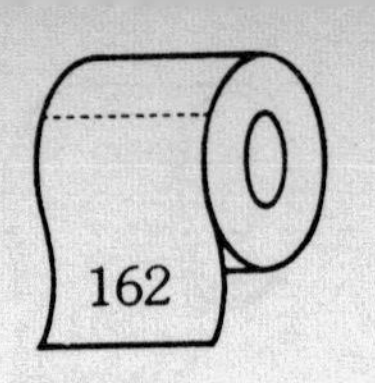

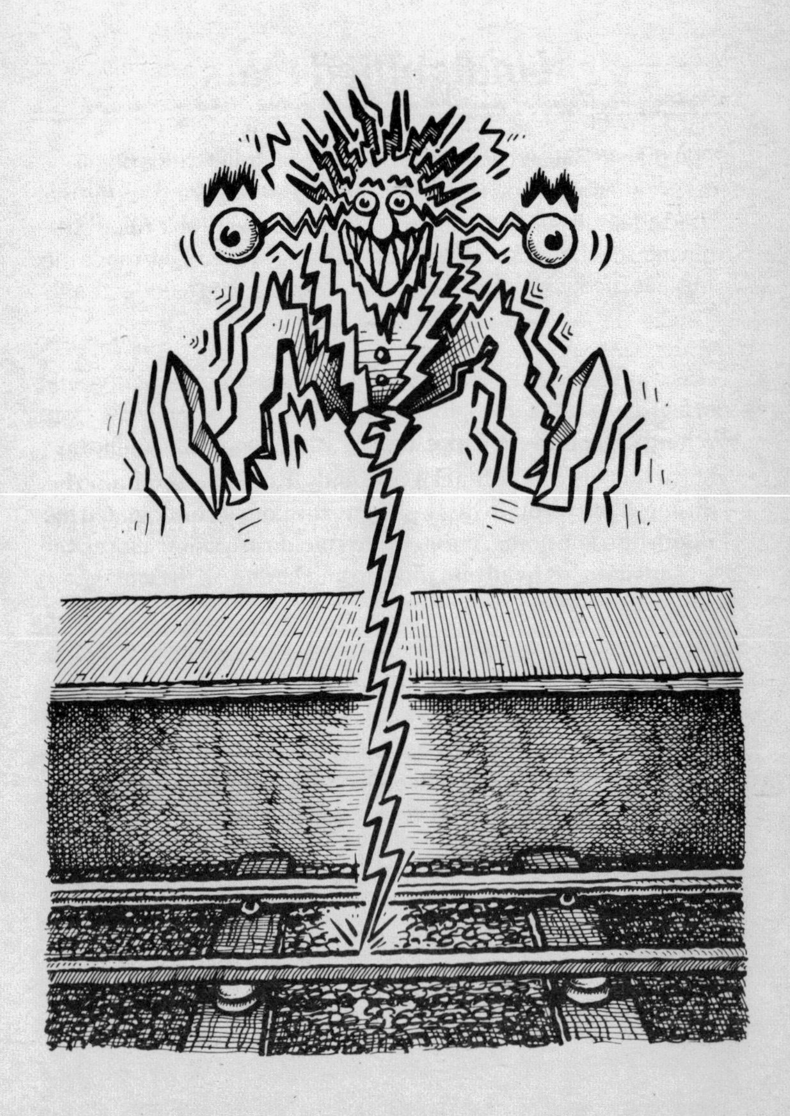

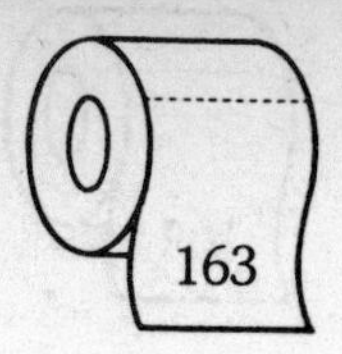

The cause of the mysterious death of Joseph Patrick O'Malley whose body was found beside the track of a New York subway was only discovered when burn marks were deciphered on his thumb, forefinger, and the tip of his penis. Drunk, O'Malley had stopped to pee beside the track. As he splashed on to the electric line, 600 volts of electricity was conducted through his urine to his body, killing him on the spot.

★ ★ ★

'A Iuwe at Tewkesbury fell into a Priue on the Saturday and would not that day bee taken out for reverence of his sabbath, wherefore Richard Clare Earle of Glocester kepte him there till Munday that he was dead.'

John Stow, *A Survey of London*, 1603.

★ ★ ★

In the public latrines in Rome, there was a bucket filled with salt water and, after using the latrines, the Romans would wipe themselves with a stick with a sponge on the end that was always left beside the bucket. According to Seneca, a German slave committed suicide by ramming the stick and sponge down his throat and choking to death.

★ ★ ★

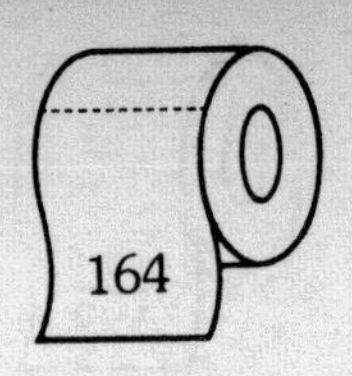

'When he came to that place that is called Constantine's Forum, a terror from a consciousness of his impieties seized Arius, which terror was accompanied by a looseness. Hereupon he inquired whether there was a House of Office near and when told there was one behind Constantine's Forum went thither. A fainting fit seized him and together with his excrement his fundament fell down forthwith and a great flux of blood followed and his small guts and blood gushed out together with his spleen and liver. He died therefore immediately.'

Socrates, *Scholastica*

★ ★ ★

There was no earl more utterly convinced of his own nobility than William Waldorf Astor, the press baron. Yet his attempts to pass himself off as a member of the British aristocracy aroused pity and contempt in London society early in the century. Even his death was somewhat unbecoming. After a heavy meal at his house, Astor withdrew to the lavatory, where he was found some time later, dead. The newspapers reported that Lord Astor had died in bed.

★ ★ ★

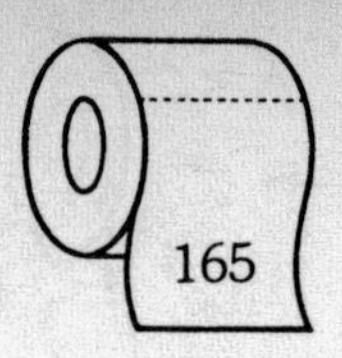

A significant natural disaster occurred in 1184 when Emperor Frederick I summoned a group of notables, including eight ruling princes, to a Diet in the Great Hall at Erfurt.

'The Emperor had occasion to go to the privy whither he was followed by some of the nobles when suddenly the floor that was under them began to sink; the Emperor immediately took hold of the iron gates of a window whereat he hung by his hands 'til some came and succoured him. Some gentlemen fell to the bottom where they perished. And it is most observable that amongst those who died was Henry, Earl of Schwatzenberg who carried presage of his death in a common appreciation of his, which was this, If I do it not, I wish I may sink in a privy.'

Reverend Nathaniel Wanley, *The Wonders of the Little World*

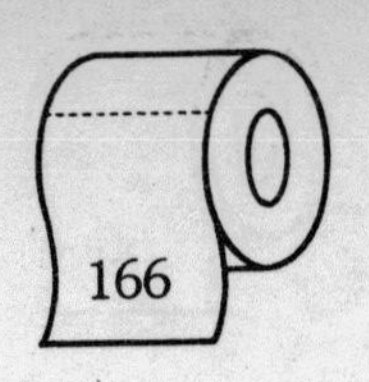

Epitaphs

Here lyeth ye body of Martin Hyde
He fell down a Midden
and grieviously dy'd.

James Hyde
his brother
fell down another.
They now lie interr'd
side by side.

Posterity will ne'er survey,
A nobler grave than this:
Here lie the bones of Castlereagh,
Stop traveller and piss.

Lord Byron

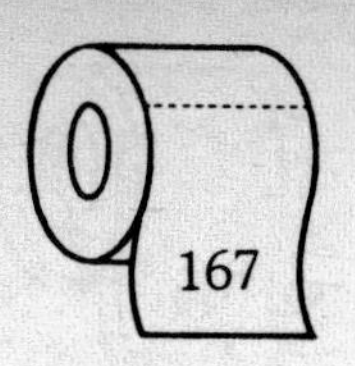

In memory of Mary Maria
wife of Wm Dodd
who died Dec. 12th
AD 1847 aged 27
also of
their children Louisa
who died Dec. 12th 1847
aged 9 months and Alfred
who died January 3rd AD 1848
aged 2 years and 9 months.

All victims to the neglect
of sanitary regulations
and specially referred to
in a recent lecture on
health in this town.

'And the Lord said to the angel that
destroyed, it is enough, stay now thy hand.' 1 Chron. XXIV

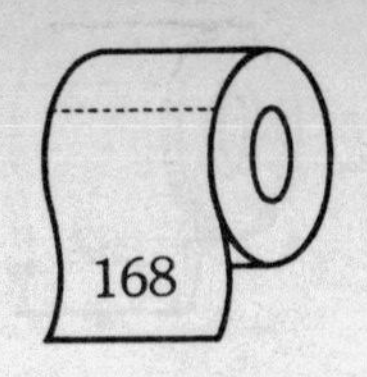

Bibliography

W. S. Baring-Gould, *The Lure of the Limerick*, Granada, 1968

Captain J. G. Bourke, *Scatologic Rites of All Nations*, Lowdermilk, 1891.

P. Bowler and J. Green, *What a Way To Go*, Pan, 1983.

Alex Comfort, *The Anxiety Makers*, Nelson, 1967.

Norman Douglas, *The Norman Douglas Limerick Book*, Anthony Blond, 1969.

Julian Franklyn, *A Dictionary of Rhyming Slang*, Routledge and Kegan Paul 1960.

Geoffrey Grigson (ed.), *The Penguin Book of Unrespectable Verse*, Penguin, 1980.

Martin Guinness, *The Shocking Book of Records*, Sphere, 1983.

Jean Harrowven, *The Limerick Makers*, Research Publishing Co., 1976.

Lucinda Lambton, *Temples of Convenience*, Gordon Fraser, 1980.

Christopher Logue, *The Bumper Book of True Stories*, Private Eye, 1980.

Morris Marples, *Public School Slang*, Constable, 1940.

Frank Muir, *A Book at Bath Time*, Heinemann, 1982.

Judith S. Neaman and Carole G. Silver, *A Dictionary of Euphemisms*, Hamish Hamilton, 1983.

Jean Nohain and F. Caradec (Trans. Warren Tute), *Le Petomane*, Souvenir Press, 1967.

Denys Parsons, *The Best of Shrdlu*, Pan, 1981.

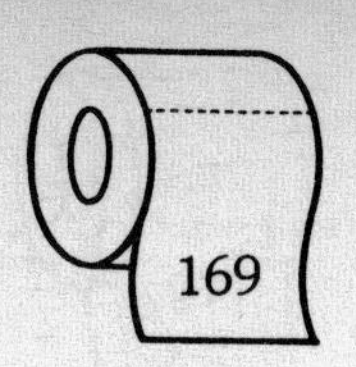

Eric Partridge (ed), *A Dictionary of Historical Slang*, Routledge and Kegan Paul, 1973.

John Pudney, *The Smallest Room*, Michael Joseph, 1954.

Punch, *Cuttings 1, 2 and 3*, Elm Tree, 1980, 1981, 1983.

Hugh Rewson, *A Dictionary of Euphemisms and Other Double-Talk*, Macdonald, 1981.

Reginald Reynolds, *Cleanliness and Godliness*, Allen and Unwin, 1946.

Jonathan Routh, *The Good Loo Guide*, Wolfe, 1965.

William Rushton, *The Filth Amendment*, Queen Anne Press, 1981.

Charles Sale, *The Specialist*, Putnam & Co., 1930.

Susan Stranks and Don Grant, *Are You Sitting Comfortably?*, Macdonald, 1980.

Wallace, Wallace, Wallechinsky and Wallace, *The Intimate Sex Lives of Famous People*, Hutchinson, 1981.

Laurence Wright, *Clean and Decent*, Routledge and Kegan Paul, 1960.

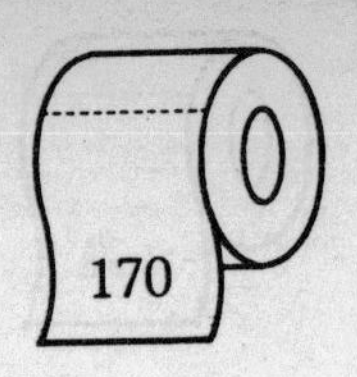

The author would like to thank the following for permission to use material in this book:
Souvenir Press for kind permission to quote from *La Petomane*; and Harold Harris for a story originally told in *The War on Land*, edited by Ronald Lewin, Hutchinson, 1969.
While we have taken every care to include the necessary acknowledgements in this book, we offer our apologies if we have inadvertently ignored any author.

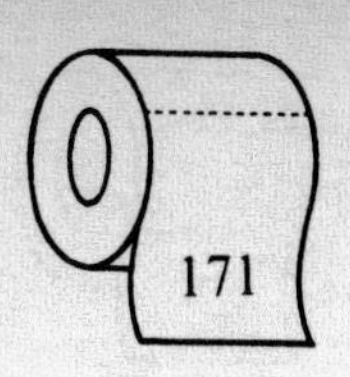

***'To keep your houses sweet, cleanse privy vaults.
To keep your souls as sweet, mend privy faults.'***

Sir John Harington, inventor of the flush lavatory, 1592